Supporting open minds since 2005

Acoustics of Materials
http://dx.doi.org/10.5772/intechopen.75245
Edited by Zine El Abiddine Fellah and Erick Ogam

Contributors
Vladimir Alshits, Dmitrii A. Bessonov, Vasilii Lyubimov, Claude Depollier, Mourad Bentahar, Marco Scalerandi, Charfeddine Mechri, Paola Antonaci, Antonio Gliozzi, Mariana Alves-Pereira, Bruce Rapley, Huub Bakker, Alice Elizabeth Gonzalez, Marcos David Fernandez, José Antonio Ballesteros, Samuel Quintana, Klara Kalinova, Ayax D. Ramirez, Stephen D. Russell, Naglaa Sami AbdelAziz

First published in London, United Kingdom, 2019 by IntechOpen
IntechOpen is the global imprint of INTECHOPEN LIMITED, registered in England and Wales, registration number: 11086078, The Shard, 25th floor, 32 London Bridge Street
London, SE19SG - United Kingdom
Printed in Croatia

British Library Cataloguing-in-Publication Data
A catalogue record for this book is available from the British Library

Additional hard copies can be obtained from orders@intechopen.com

Acoustics of Materials
Edited by Zine El Abiddine Fellah and Erick Ogam
p. cm.
Print ISBN 978-1-83880-349-0
Online ISBN 978-1-83880-350-6

For EU product safety concerns:
IN TECH d.o.o., Prolaz Marije Krucifikse Kozulić 3, 51000 Rijeka, Croatia, info@intechopen.com or visit our website at intechopen.com.

Meet the editors

Zine El Abiddine Fellah graduated from the University Science and Technology Houari Boumedienne, Algeria, in Physics (1996). He obtained a Ph.D. degree in Acoustics at the University of Le Mans in France (2000). He has worked at the Acoustic and Thermal Laboratory of Leuven, Belgium (2001), and at the Therapeutic Ultrasound Research Laboratory of the National Institute of Health and Medical Research, Lyon, France (2002–2004). He is currently a research scientist at the Centre National de la Recherche Scientifique at Marseille, France. He received the Yves Rocard Prize in 2003 from the French Acoustical Society. His research interests are in the area of ultrasound, low-frequency measurements, direct and inverse problems, fractional calculus, porous media, fractals, micropolar media, and cancellous bone.

Erick Ogam pursued his secondary and higher education in Kenya and France, respectively. He received his PhD degree from the Université de Provence Aix-Marseille I, France, in 2007. He is currently a research engineer at the Laboratoire de Mécanique et d'Acoustique, UMR7031 CNRS, Marseille, France. His research interests include direct and inverse problems in acoustics and vibroacoustics, modeling ultrasonic wave propagation in complex media, fluid structure interaction, non-linear vibration, developing acoustic and vibration transducers/sensors, data acquisition, generating real data, new experimental techniques, and signal processing.

Contents

Preface

Materials in their different forms and compositions are ubiquitous in our environment. Soil and rocks are the most common natural examples, whose characterization is of prime interest in many fields (geology, the oil industry, etc.). Building materials (like concrete and road surfaces), the so-called acoustic materials (plastic foams, fibrous maerials), and many others used in transport, building, infrastructures, etc., are today used in the field of noise reduction (over the last 30 years, acoustic standards have become increasingly stringent, making the fight against noise pollution a key economic factor). The materials used in structures and microstructures are now subject to increasingly stringent non-destructive controls and the demand for increased performance of these controls remains fully on the agenda. Finally, biomaterials, living tissues, bone tissues (cancellous bone, lungs, etc.) are the subject of increasingly sophisticated analyses for the diagnosis of certain pathologies such as osteoporosis, which require precision materials. Thus, whether it concerns seismic studies, environmental acoustics, non-destructive testing, or medical diagnostics, the application of acoustic techniques applied to materials still largely opens the way to fundamental studies on their propagation in complex media, the modeling of their characteristics and defects, experimental techniques used to characterize them, and the means to provide for their transfer to applications. In a broad sense, acoustic waves are an effective means for the modeling and characterization of materials.

Acoustic waves have the ability to probe materials during propagation. The interactions between wave and medium highlight the physical properties intrinsic to materials, thus allowing the characterization of their mechanical properties. In this book, several authors with extensive industrial and academic experience have helped to showcase the potential applications of acoustic waves on materials. Considerable progress has been made in materials acoustics. High-performance theoretical models and elaborate experimental designs have been developed with the progress of sensors, allowing for better theoretical predictions and experimental characterizations. This book is divided into three chapters covering applications ranging from sound absorption (polymeric nanofibers) to propagation in complex media (porous biological structures, acoustics for interiors) and non-destructive control (nonlinearity, crystals, leakage). The book is intended to be an ideal reference for a wide audience interested in acoustic propagation in materials with various applications.

For their frankness and diligence in reviewing the proposed chapters, we would like to thank all the authors who have contributed to this book. We have incorporated many of their recommendations and the book is much improved as a consequence.

Special thanks go to Anja Filipović (Commissioning Editor), Manuela Gabrić and Dolores Kuzelj (Author Service Managers) for their dedicated support in the reviewing process and their suggestions for further improvements. Finally, all thanks to IntechOpen for publishing this book

Zine El Abiddine Fellah and Erick Ogam
Aix-Marseille University,
Centrale Marseille,
Marseille Cedex,
France

Section 1

Control of Sound - Absorbing Materials for Damping of Sound

Chapter 1

How Do Acoustic Materials Work?

Alice Elizabeth González

Abstract

Acoustic quality of closed spaces is an increasing concern all around the world, since noise pollution is one of the main nowadays pollutants, but also one of the less considered when building designing and construction. In 2011, the World Health Organization stated that noise pollution should be treated as a public health concern: about 1 million years of human healthy life are lost yearly because of the environmental noise pollution, emphasizing on traffic noise in the cities. There are some physics phenomena that are the rule of thumb for room acoustic projects. This chapter introduces the main concepts about them: sound absorption, insulation, and diffusion. Their principles, main implementation, and computing are presented for each one.

Keywords: acoustic materials, sound absorption, sound insulation, sound diffusion

1. Introduction

1.1 Acoustic quality

Under the term "acoustic quality," a set of characteristics (as sound pressure levels, spectral composition, and duration of the sounds perceived) is integrated, as well as others related to the space itself—for example, its reverberation time—which allow to qualify how valuable this space is regarding its aptitudes or potentialities for the desired use.

One place can have good acoustic quality for a certain use but not for another. For example, the high reverberation time of Catholic churches is part of the characteristics of the space of meditation that is desired to be generated there and is suitable for interpreting/listening to sacred music; however, it conspires against the understanding of the spoken word.

1.2 Noise control and acoustic project

The concepts of acoustic quality and noise control are often closely related. Noise control refers to a set of methods, techniques, and technologies that allow obtaining acceptable noise levels in a certain place, according to economic and operational considerations [1].

Noise control does not necessarily imply reduction of noise emissions; it refers to making acceptable the sound level in immission (i.e., the signal that reaches the receiver). To know if it is, some objectives and valid criteria must be selected and applied to compare with, in order to answer the question of "acceptable for what" or "for whom". There are different ways to attack a wide range of cases in order to achieve the desired acoustic quality at the receiver.

The acoustic project of enclosures involves the selection of materials to determine the type and quality of walls, surfaces, etc. so that a certain location is apt for one use. It involves avoiding an undesired level of incidence of external noise and making the internal reverberation characteristics adequate for the desired use. To achieve the acoustic quality objectives, working harmonically on insulation, absorption, and diffusion of sound is needed.

1.3 Sound insulation and absorption

When a sound wave reaches one surface, part of its energy (incident energy E_i) is reflected toward the same half space from where it comes (reflected energy E_r). According to Snell's law, the angles of E_i and E_r with the surface of incidence are equal. When the propagation media changes from media 1 to media 2, the incident and refracted angles should fulfill next relation (also according to Snell's law):

$$n_1 \sin\theta_1 = n_2 \sin\theta_2 \tag{1}$$

where n_1 and n_2 are the relations with sound speed c_1 and c_2 in the considered media and they fulfill the relation (c_0 is the reference sound speed in air):

$$c_0 = n_1\, c_1 = n_2\, c_2 \tag{2}$$

The non-reflected energy is usually expressed as the sum of two terms: transmitted energy E_t and absorbed energy E_a. The first one is the part of the energy that passes through the surface or wall and generates another acoustic wave at the other side of the wall; the second one is the part of the incident energy that is dissipated at the surface (**Figure 1**).

Let α be the absorption coefficient ($\alpha = E_a/E_i$) and τ the transmission coefficient ($\tau = E_t/E_i$). The part of the incident energy that is not reflected should accomplish:

$$E_i - E_r = \alpha E_i + \tau\, E_i = (\alpha + \tau)\, E_i \tag{3}$$

The more energy is absorbed, the less energy should be transmitted and vice versa. In other words, a good insulating material is bad for acoustic absorption, as a high value of α is related to a low value of τ. The more porous is a material, the less it insulates; the more resistance to the flow of air a material presents, the better insulation performance it has. Intuitively (but not strictly), the heavier the material is, the better insulation performance is expected.

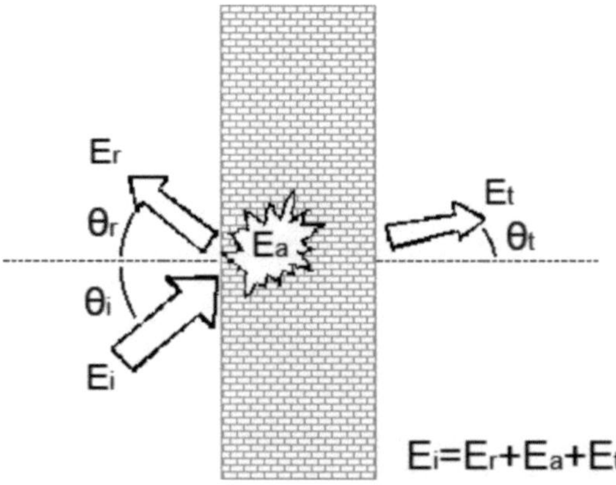

Figure 1.
Possible destinations of the acoustic energy that reaches a surface.

When a good insulation is reached, a great amount of the incident energy is retained in the emission source room. If the room has low sound absorption, the sound pressure levels inside could eventually increase. To avoid this result, a suitable absorption treatment should be done on the surfaces of the room in order to reduce the reflections (the reverberation).

The so-called absorbed energy is dissipated in the surface on which the sound waves impinge. It is related to the characteristics of the surface material, both in terms of its internal structure and its elasticity and texture: the more elastic or rough the surface is, the more energy will be absorbed, as it will deform more or the sound path will be increased the more through multiple reflections. The dissipation that occurs in the pores and the microstructure of the material will be higher, and less energy will be reflected toward the room. Anyway, the amount of energy involved in these phenomena is very little to cause a perceptible change of the surface temperature or shape.

However, if acoustic energy is emitted in a low absorption room, the sound pressure levels inside can be increased by successive sound reflections. Consequently, the insulation required to protect the contiguous rooms against sound transmission may be higher.

Regarding what happens at the emission room, the acoustic energy distribution is strongly related to its acoustic quality. If a homogeneous distribution is achieved, every people at the room will have the same sound quality experience. In fact, there are some common acoustic defects that can occur, for example, stationary waves due to normal modes. To work on the acoustic energy distribution into a room, attention on sound diffusion is needed. Sound diffusion materials are those that contribute to scatter sound waves in different directions to reach a homogeneous (or diffuse) sound field. The main property of sound diffusion materials is their surface design: they have irregular surfaces with cavities and protuberances whose dimensions are calculated according to the sound frequencies they are expected to correct.

2. Sound absorption

The acoustic absorption is the phenomenon by which the acoustic energy is transformed into another type of energy: thermal, mechanical, or deformation energy. The acoustic absorption is, then, an energy dissipation phenomenon.

The acoustic absorption coefficient α of a surface is defined as the relation between the acoustic energy that it can absorb and the incident energy that affects it. It is dimensionless:

$$\alpha = \frac{E_a}{E_i} \tag{4}$$

Then, the absorbed energy turns out to be $E_a = \alpha E_i$.

There are three families of absorbent materials, which fulfill their function according to different phenomena: porous/fibrous materials, membrane absorbers, and resonators. Each one has its best performance in different frequency ranges, as shown in **Figure 2**.

2.1 Porous or fibrous sound absorbers

Absorbent materials are usually elastic, not very dense and permeable; in fact, they are formed mostly by air. These are soft or fibrous materials containing fine

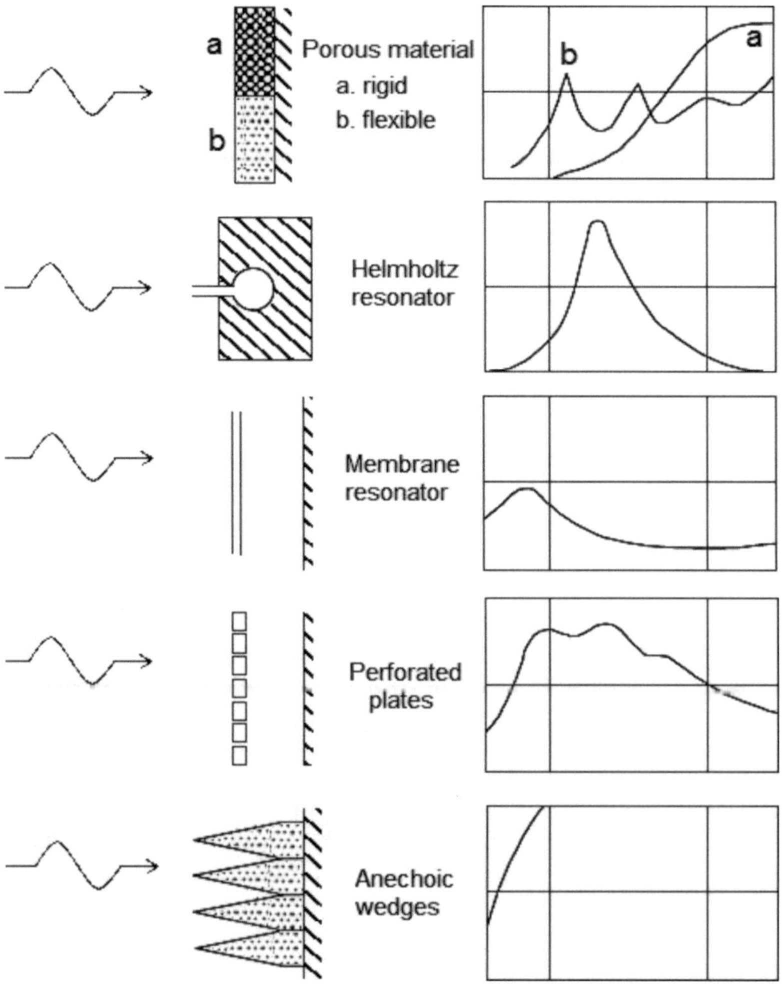

Figure 2.
Types of acoustic absorbers and their performance curves. Adapted from [2].

channels interconnected with each other. Although they are considered the absorbent materials par excellence, they are not the only ones. They can absorb acoustic energy through two mechanisms:

- When they are soft materials, they absorb due to the deformation that occurs when the sound wave hits them.
- When they are porous materials, they absorb by the vibration of the air contained in its pores, which loses energy by friction against their edges.

The performance of these absorbers improves the smaller the length of the incident wave in relation to the dimensions of the irregularities of their surfaces. Therefore, their performance improves with increasing frequency, and it is usually at least good at most conversational frequencies. Because of their high air content, their acoustic impedance Z is very close to that of the air (Z_{air}). Then, most of the energy of the incident wave will tend to penetrate the material, and only a small fraction will be reflected.

The fibrous absorbent materials par excellence are glass wool and rock wool. However, the performance of many materials, everyday objects, and even people in different conditions has been studied from the point of view of their behavior as acoustic absorbers.

It should be taken into account that for a fibrous absorbent material to work well, its pores must not be clogged. When they are mounted in a polluted atmosphere place, their performance will decline if the pores become saturated with particles. Sometimes to protect them, they are covered with fabric or even with a grid or perforated plate, although strictly some acoustically active surface is lost. They should not be covered with rigid materials that hide them completely, since it is necessary that the porous/fibrous surface be kept available to absorb the sound.

The thickness of an absorbent material is very important for its performance: the greater the thickness, the more the opportunities to lose energy for the incident wave.

If the absorbent material had an infinite thickness, it could be considered a perfect absorber, and all the energy of the incident wave could be absorbed. But the actual thickness is limited, and there is usually a reflecting surface behind the absorber. The sound waves will be reflected in this material, but they will have to cross twice (round trip) the thickness of the acoustic absorbent to return to the emitting room. They suffer a lot of reflections to get through, and they lose energy in each of them, mainly by friction or deformation. These energy loses are those that, in short, "spend" the acoustic energy and reduce the amplitude of the reflected wave that returns to the emitting environment.

In porous or fibrous absorbent materials, it is valid to assume that the sound pressure level decays linearly with the thickness.

An empirical result is that the best performance is given for a material thickness equal to or greater than a quarter of the wavelength ($\lambda/4$), so these materials will be effective for medium to high frequencies. For example, for a frequency of 1.000 Hz, the corresponding wavelength is 34 cm, and then $\lambda/4$ is 8.5 cm, which would be a reasonable thickness of fibrous absorbent material to be placed. Sometimes, when it is desired to extend the operating range, faceted foams or anechoic wedges are placed, which may have surface irregularities of several centimeters in height (15 cm and even more).

To improve performance at low frequencies, an air chamber can be left between the absorbent material and the rigid face to be treated. Although in theory the minimum distance between the absorber and the facing should also be at least $\lambda/4$ of the main frequency to be absorbed, it is empirically recommended that the separation be at least $\lambda/10$.

As the thickness of the air chamber increases, the best performance of the material moves to lower frequencies. The same occurs when a protection is applied over the absorbent material.

The density of the material improves the performance as it increases to an optimal point, but if it continues to increase even more, the material will gain rigidity, and its performance will worsen, as it will start to perform more like a solid than a fibrous material (it will no longer be composed mostly of air). In general, it is assumed that the density of a fibrous sound absorbent material should not exceed 100 kg/m^3.

There are many ways to mount the absorbent materials on different surfaces (walls, ceilings, floors); they can also be installed as suspended panels. The value of α could vary not only with the frequency of the incident wave but also with the way the material is installed.

Figure 3 shows acoustic absorbers made with waste from the textile industry (weaving). These panels were made with agglomerated wool remnants that were then surface-patterned to improve their absorption and diffusion characteristics. They are low-cost panels developed at the University of the Republic (School of

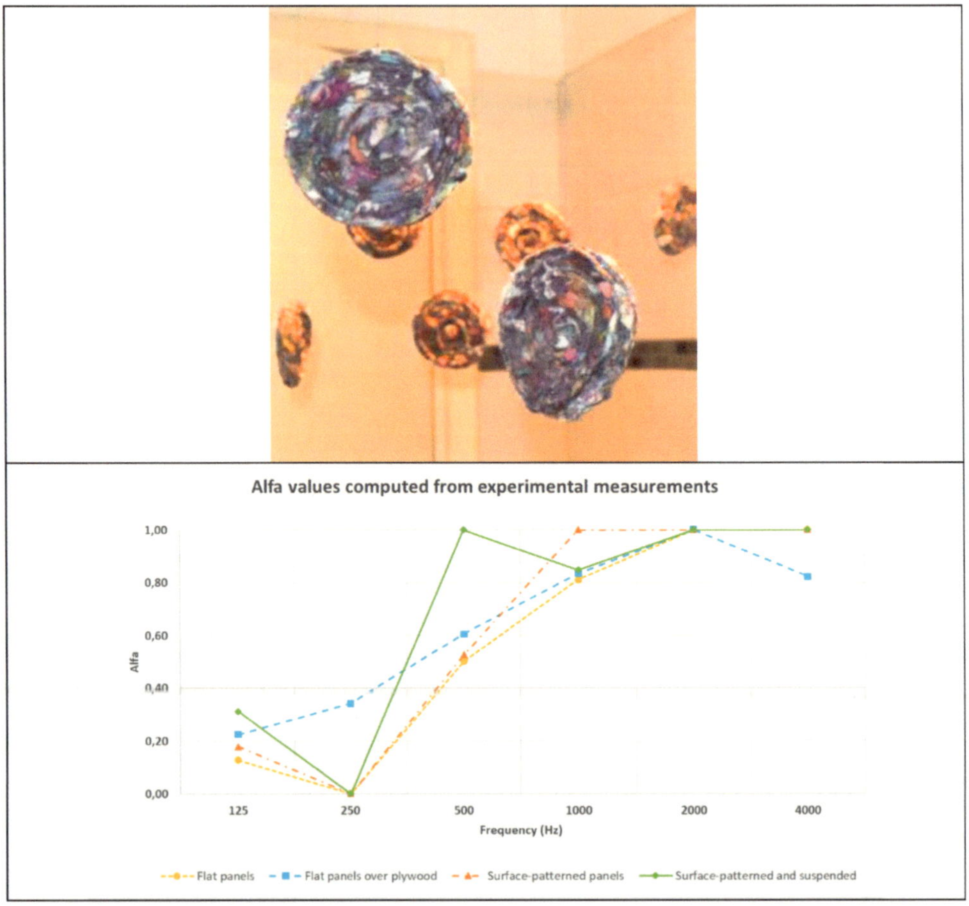

Figure 3.
Acoustic absorbers made with weaving wastes. Up: suspended mounting. Down: experimental absorption coefficient for different ways of mounting.

Design and Faculty of Engineering) that demonstrated good acoustic performance and very low manufacturing cost, in addition to allowing the recycling of a very common waste in Uruguay [3].

2.2 Membrane or plate absorbers

A membrane or plate absorber is an air impervious material fixed at its edges at a certain distance from a rigid surface, so as to leave a tight cavity between both (**Figure 4**). When the stiffness of the material is negligible in relation to the tensions that hold it, it is said to be a membrane; when its stiffness must be considered, the material is said to be a plate.

The best performance of the plate is expected to occur at its resonance frequency and can be calculated as

$$f_r = \frac{60}{\sqrt{m\,d}} \tag{5}$$

where m is the surface mass of the plate, in kg/m^2; and d is the distance from the panel.

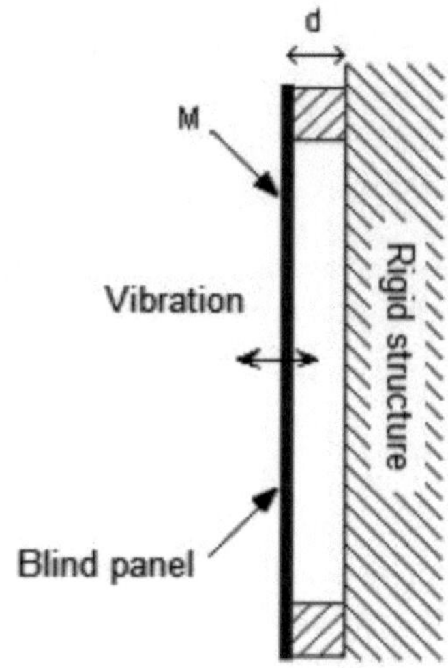

Figure 4.
Typical acoustic absorbent panel.

In any case, the value of α coefficient of these plates is usually not more than 0.50. Since the surface mass of the panels has a practical minimum, these absorbers are useful only at low frequencies. Increasing the mass *m* of the panel or the distance *d* lowers the resonance frequency. If the cavity is filled with a soft material, the operating range broadens in terms of frequencies, but in turn it lowers the maximum value of α of the plate.

2.3 Resonator absorbers

The acoustic resonators consist of a cavity that communicates with the outside by a narrower conduit or neck, and of such dimensions they can dissipate energy in a certain frequency, that is, its resonance frequency (**Figure 5**).

The resonance frequency of a Helmholtz-type resonator can be calculated as follows:

$$f_{res} = \frac{c}{2\pi} \cdot \sqrt{\frac{S}{V\, l_e}} \tag{6}$$

where S being the neck area of the resonator, in m^2; V the inside volume of the resonator, in m^3; and l is the effective length of the neck of the resonator, in m: $l_e = l_{real} + 1.7\, r$; r the radius of the neck, in m.

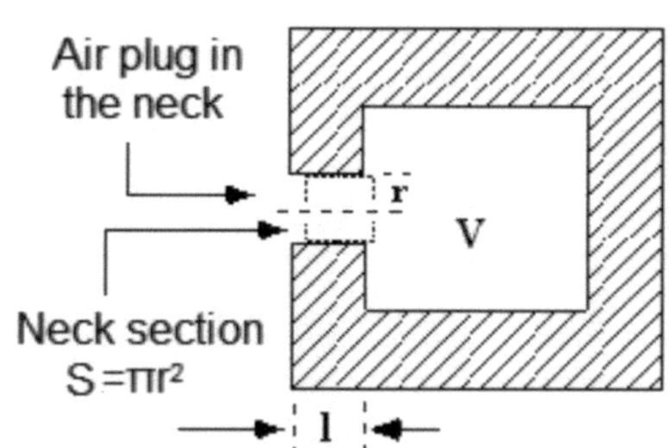

Figure 5.
Outline of a Helmholtz acoustic resonator (adapted from [1]).

As these devices are very selective, they are generally used only when a reverberation has to be eliminated at a well-defined frequency or, in any case, in a fairly narrow frequency range. If some sound absorbent material is placed inside the cavity and especially close to the neck, the range of frequencies absorbed by the resonator can be expanded somewhat, but the efficiency in the frequency of better performance will decrease.

The construction of Helmholtz resonators is usually done by panels with circular or linear perforations, where the total area of the resonator is the sum of the areas of all the holes.

3. Sound insulation

Airborne sound propagation is usually due to the elastic vibrations of the air due to the sound waves that reach a material surface and excites it. It also occurs, more easily, through evident discontinuities (e.g., openings) or unwanted weak spots (such as cracks, weakly sealed passages for electrical, sanitary or other installations, ventilation ducts, and openings that do not close properly or whose frames are poorly sealed, among other possible imperfections).

3.1 How does sound insulation occur?

The acoustic insulation of a material or a set of them refers to the property to oppose and, consequently, to reduce the flow of acoustic energy that passes through it. A good acoustic insulation reduces the acoustic energy that is received on the other side of it, affecting the amplitude of the wave but not through dissipative phenomena, as the acoustic absorbers do.

The acoustic insulation is based on the modification of the amplitude of a wave when passing from a medium of acoustic impedance Z_1 to another of acoustic impedance Z_2. The relationship between the acoustic impedances does not only have to do with the modification of the amplitude of the incident wave: it is also related to the fraction of the incident energy that will be reflected or transmitted. Most of the acoustic properties of materials can be studied by the application of new laboratory and numerical methods as ultrasonic characterization [4], inverse characterization with basis on the acoustic impedance measurement [5], or ensemble averaged scattering [6].

The factors of reflection and transmission can be written according to the values of the impedances of the two media involved:

$$F_r = \left(\frac{Z_1 - Z_2}{Z_1 + Z_2}\right)^2, \qquad F_t = \frac{4Z_1Z_2}{(Z_1 + Z_2)^2} \tag{7}$$

It can be seen that for similar values of Z_1 and Z_2, the reflection coefficient F_r tends to zero, and the more different their values are, the value of F_t tends to zero.

Then, the greater the difference between the values of Z_1 and Z_2, the greater the fraction of the energy of the incident wave that will be reflected, and, consequently, the non-reflected energy, which comprises the energy absorbed and the energy transmitted, will be lower. Implicitly, even if the absorption were not significant, a large difference between the impedances of the two media tends to reduce the amount of acoustic energy transmitted. Taking into account both premises at the same time, when a wave propagates by air and reaches a material surface, most of the energy will be reflected, and only a small portion will be transmitted to the wall, as its acoustic impedance is undoubtedly much greater than that of air. When passing from the wall again to the air, most of the energy will be again reflected inside the wall, and only a small portion will be transmitted to the air. Then, the transmitted wave will have a smaller amplitude than the wave that would result if the propagation media has not changed (**Figure 6**).

The transmission coefficient τ of a material is then the relation between the transmitted energy and the incident energy:

$$\tau = \frac{E_t}{E_i} \tag{8}$$

The acoustic reduction index R is defined with basis on τ:

$$R = 10 \log \frac{1}{\tau} \tag{9}$$

3.2 Acoustic performance of a single wall

A single wall in acoustics is formed by only one foil. If it is a macroscopically homogeneous wall, its acoustic insulation will depend on several of its mechanical properties.

Intuitively, a simple and homogeneous wall offers good sound insulation when it is heavy and tight to the passage of air but only weakly rigid. A more rigorous analysis allows recognizing several zones with different behavior, as shown in **Figure 7**: the design zones are those controlled by mass or by coincidence; the zones controlled by stiffness or resonance refer to a poor and irregular acoustic performance.

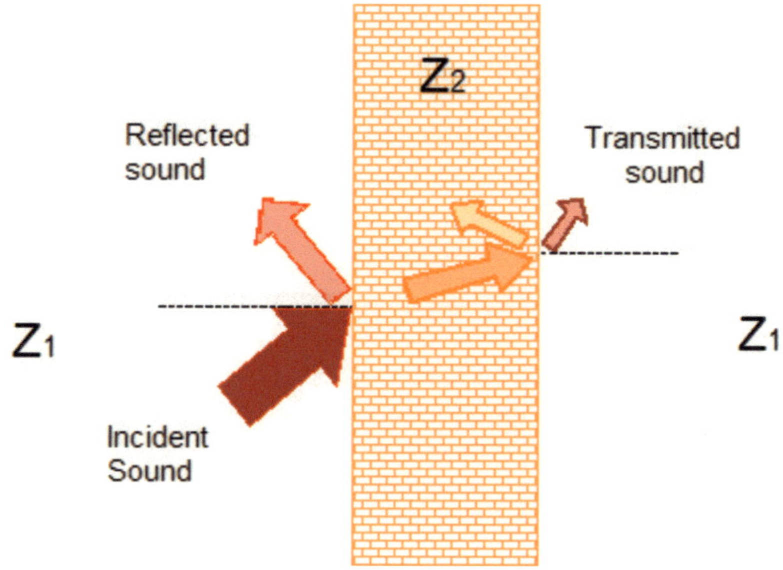

Figure 6.
Destinations of the acoustic energy that reaches a wall.

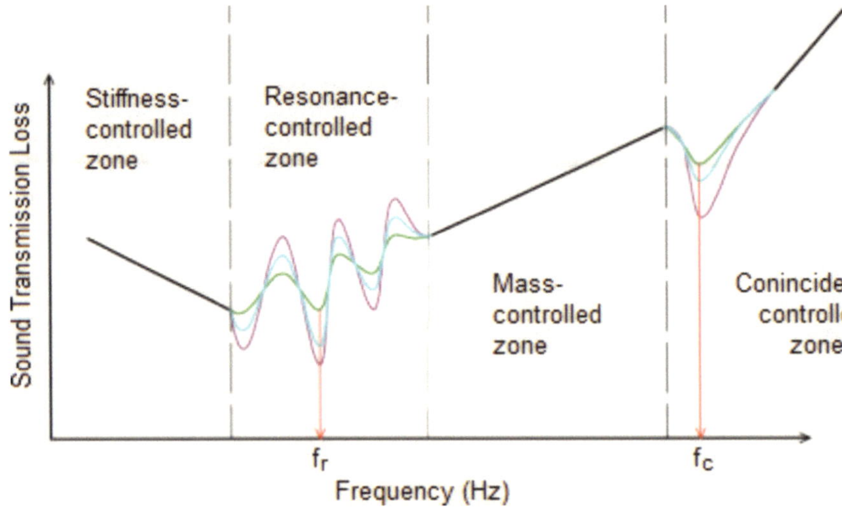

Figure 7.
Acoustic performance of a simple homogeneous wall (adapted from [7]).

1. **Stiffness-controlled zone:** it corresponds to the lowest frequencies, below the frequency of resonance f_r. The greater the stiffness k[1], the poorer the insulation of the wall. The resonance frequency f_{res} can be calculated as:

$$f_{res} = \frac{1}{2\pi}\sqrt{\frac{E\,e}{m\,S}} \tag{10}$$

where m is the surface density[2] of the material, kg/m^2; e the thickness of the wall, in m; S the area of the wall, in m^2; and E is the Young modulus of the material, in N/m^2.

2. **Resonance-controlled zone:** it occurs between f_{res} and $2f_{res}$. In this zone, the behavior is irregular and unpredictable; points of very poor performance are alternated with others that are quite good.

3. **Mass-controlled zone:** it is the preferred zone for acoustic design. It covers from $2f_{res}$ to half of the critical or coincidence frequency f_c. The insulation depends on the surface mass and the frequency of the incident wave. It is usually computed according to the so-called masses law. An improvement of insulation of about 5.4–6 dB is expected each time frequency or mass duplicates. Close to the extremes of this zone, the wall insulation performance is rather poor.

   The acoustic insulation performance of a simple homogeneous wall in the mass-controlled zone can be computed by the well-known expression of the masses law [4]:

$$\text{TL} = 20\ \log m + 20\ \log f - 42 \tag{11}$$

   where m is the surface mass in kg/m^2 and f is the central frequency of the octave band in Hz.

---

[1] The stiffness k is the ability of a solid to withstand stresses without acquiring deformations. A wall can be idealized as a thin plate, so its flexural stiffness (i.e., the stress generated by a load perpendicular to it) can be expressed as $k = \frac{E}{1-\nu^2} \times \frac{e^3}{12}$, where ν is Poisson's coefficient of the material.

[2] The surface density or surface mass of a material is the mass of 1 m^2 of it with its current thickness. Otherwise, $\boldsymbol{m}$ is the product of the volumetric density $\boldsymbol{\rho}$ of the material by its thickness $\boldsymbol{e}$.

A more conservative expression is proposed by [1] to take into account the irregularities and imperfections of the real materials:

$$\mathrm{TL} = 18\ \log m + 18\ \log f - 45 \tag{12}$$

4. **Coincidence-controlled zone:** it is the zone over the mass-controlled one. In general, the wall performance is similar to that at the mass-controlled one except when frequency is close to the coincidence or critical frequency f_c. At this frequency, the bending waves that propagate in the material can coincide with the sound waves that propagate by the air, generating high amplitude vibrations in the wall. This phenomenon can occur above the so-called critical frequency f_c, which is the one at which the frequency of the incident waves coincides with that of the longitudinal waves of bending of the wall. The insulation weakens, because the acoustic energy is transmitted through the divider in the form of bending waves, coupled with the acoustic waves in the air. This frequency depends exclusively on the material of the wall and its thickness. The critical frequency is calculated as

$$f_c = \frac{c^2}{\pi}\sqrt{\frac{3m}{E e^3}} \tag{13}$$

where c is the speed of sound propagation in air, in m/s; e the thickness of the wall, in m; m the surface density, in kg/m^2; and E is the Young modulus of the material, in N/m^2.

From $f_c/2$ onwards, the phenomenon of coincidence can occur. Then, $f_c/2$ is often assumed as the limit of validity of the masses law. For frequencies higher than f_c, the acoustic behavior is governed by the internal damping of the material; the insulation grows again from the value corresponding to f_c. In this area, the law of mass is still used, although it is known that there is a very important drop in a frequency close to the critical frequency.

If the critical frequency is expressed as a function of the resonance frequency, they turn out to be inversely proportional, so when selecting a partition with a low resonance frequency, it also has a high critical frequency:

$$f_c = 0.0877\ \frac{c^2}{e f_{res}\ \sqrt{S}} \tag{14}$$

S is the area of the partition (m^2).

The critical frequency is also inversely proportional to the stiffness k of the partition. The greater the stiffness is, the lower its critical frequency (and the higher its resonance frequency):

$$f_c = 0.159 c^2 \sqrt{\frac{m}{k\ (1-\upsilon^2)}} \tag{15}$$

Table 1 presents the values of the Young and Poisson modules, the volumetric densities, and the critical frequencies for various materials, for a thickness of 1 cm. To obtain the critical frequency f_c for other thicknesses, dividing the value of the table between the new thicknesses in cm is needed.

3.3 Double walls

If the mass of an insulating element is doubled, the acoustic insulation improvement that can be achieved will be of a theoretical maximum of 6 dB, according to

	Young modulus E, $N/m^2 \times 10^{10}$	Poisson modulus ν, $N/m^2 \times 10^{10}$	Density (kg/m^3)	Critical frequency for 1 cm thickness (Hz)
Solid bricks	2.50×10^{10}	–	2000	2700
Reinforced concrete	2.61×10^{10}	–	2600	1900
Steel	1.95×10^{11}	0.31	7800	1300
Aluminum	7.16×10^{10}	0.4	2700	1200
Lead	1.58×10^{10}	0.43	11,300	5500
Asbestos cement	1.50×10^{10}	–	1090	1700
Gypsum	4.69×10^{9}	–	1150	2700
Plasterboard	–	–	875	4600
Glass	6.76×10^{10}	0.22	2500	1200
Pinus wood	1.40×10^{10}	0.18	700	1700
Agglomerated wood	7×10^{9}	–	750	2700
Plywood	–	–	600	2100
Cork	5×10^{6}	0.28	250	18,000
Rubber	7×10^{6}	0.4	1100	85,000
Extruded polystyrene	–	–	33	10,900

Table 1.
Density and critical frequency of some materials (from different sources).

masses law. But if the mass increase is done by distributing it in two not linked elements separated by an air chamber, the performance of the whole is better than that which results simply from doubling the mass of the original element. This is due to the changes in acoustic impedance to which the sound waves are undergoing when passing through the different elements of the set. The acoustic performance is further improved if the two elements are not exactly the same, and it can be even better if a sound absorbent material is placed into the air chamber.

It is very important to meet a real constructive independence between both foils of the wall, since otherwise the whole will not work as a double wall but simply as a wall whose mass is the sum of the masses of the two elements. Distributing nonhomogeneously the total mass is advantageous, since it allows to achieve that both partitions have different critical and resonance frequencies, avoiding the occurrence of frequencies for which the overall performance would be very poor. If the two wall foils are rigidly joined, the insulation of the whole will decrease. If a reasonably independence is constructively achieved (i.e., they do not vibrate together), the resonance frequency of the overall turns out to be

$$f_r = 60\sqrt{\frac{1}{d}\left(\frac{1}{m_1}+\frac{1}{m_2}\right)} \qquad (16)$$

where m_1 and m_2 are the surface masses of both elements, in kg/m^2; and d is the thickness of the air chamber or separation between partitions, in m.

Double walls have a good acoustic performance for the frequencies between their resonance frequency and their critical frequency.

The resonance frequency will be lower the greater the masses of the two foils and/or the greater the distance between them. When the air chamber between the two foils of the double wall is filled with absorbent material, the resonance frequency of the whole decreases (about 85 % of the calculated value). The desirable values for f_{res} are less than 100 Hz and as an optimum condition, less than 60 Hz. Close to the resonance frequency, the insulation is very poor; hence, it should be lower than the most probable minimum incidence frequencies (for human speech they can typically be up to 80 Hz and for music, even lower than 40 Hz). For frequencies below f_{res}, the behavior of the set of the double wall is, at most, that of a simple wall with a surface mass equal to (m_1+m_2), although its performance could be poorer.

The end of the range in which the double wall has its best performance occurs at its critical frequency f_c, which does not depend on the masses of the partitions but only on the distance d between them:

$$f_c = \frac{c}{2d} \tag{17}$$

Every time d equals a whole number of half wavelengths, the acoustic insulation has an important decay. This occurs at the harmonic frequencies of f_c. For these frequencies, stationary waves of $n.f_c$ frequencies (n = 1, 2, 3, ...) occur in the air chamber, which further weaken the insulation.

Stationary waves are periodic waves where the nodes and crests occupy fixed positions that do not vary in time. Because d is an integer multiple of $\lambda/2$, a node will occur on the second foil of the double wall, and the wave will be reflected with very little energy dissipation, with at least one relative maximum remaining in the air chamber. To attenuate this deleterious effect, absorbent material can be placed on one side of the cavity. This not only improves the insulation of the set by dissipating acoustic energy but also improves its acoustic performance by lowering its resonance frequency.

As practical criteria, if the double partition is composed of two light, flexible sheets of surface masses m_1 and m_2 expressed in kg/m^2, then the distance d between sheets must comply with

$$d\ [m] \geq \left(\frac{1}{m_1} + \frac{1}{m_2}\right) \tag{18}$$

Also, the air chamber might contain a nonrigid porous absorbent material.

Predicting the acoustic insulation of a double wall is not obvious, because of the different phenomena involved in its performance. Referring to laboratory tests is hardly recommended. **Table 2** shows some results from laboratory tests of double walls.

Sometimes not only double walls are used but also three-foiled walls. The involved acoustic phenomena are so complex that laboratory testing is mandatory to determine their acoustic insulation performance.

3.4 Some causes that weaken the acoustic performance of a wall

Although when the calculations are rightly performed, there are some factors that can make the forecasts much more optimistic than the actual acoustic

Description	m(kg/m²)	Frequencies (Hz)					
		125	250	500	1000	2000	4000
15 cm brick wall with plaster, plus 25 mm glass wool plus 4 cm air chamber plus 25 mm glass wool plus 7 cm concrete	350	40	54	57	65	70	76
30 cm brick wall plus 50 mm glass wool plus 12 cm air chamber plus 30 cm bricks with plaster	s/d	80	90	98	109	111	111
Two 6 mm plywood panels, each one glued to both faces of wood crossbars of 2.5 cm x 7.5 cm and 41 cm separation	12.2	16	18	26	28	37	33
Two 6 mm plywood panels, each one glued to both faces of wood crossbars of 2.5 cm x 7.5 cm and 41 cm separation plus 10 cm air chamber	14.1	14	20	28	33	40	50
Two 5 mm plywood panels with 1.5 mm lead foil between them	25	26	30	34	38	42	44
7 cm light concrete plus 1.5 cm plaster, 20 mm glass wool, 3 cm air chamber, 5 cm light concrete plus 1.5 cm plaster	450	24	33	41	50	60	65
20 cm concrete plus 25 mm glass wool plus 12 cm air chamber plus 25 mm glass wool plus 15 cm brick wall	650	63	72	74	85	91	93
9 cm concrete plus 2.5 cm air chamber plus 65 mm rigid glass wool panel plus 9 cm concrete plus 16 mm gypsum plate	s/d	49	54	57	56	71	81
9 cm concrete plus 6 cm air chamber plus 65 mm rigid glass wool panel plus 9 cm concrete plus 16 mm gypsum plate	s/d	57	65	76	82	86	83
Two 12 mm gypsum plates with a 7 cm air chamber between them	s/d	13	21	33	43	44	39
One gypsum plate 2 x 12 mm and one 12 mm gypsum plate with 7 cm air chamber between them	s/d	18	25	39	47	49	44
Two 2 x 12 mm gypsum plates with a 7 cm air chamber between them	s/d	23	30	45	49	52	52
Two 12 mm gypsum plates with a 2 cm air chamber and 50 mm glass wool between them	s/d	21	35	48	55	56	43

Table 2.
Acoustic insulation of some tested double walls (values from [8–10]).

performance of a wall. Two of the main causes of this are the presence of weak points or cavities and the contributions by nondirect transmission (lateral and/or solid), which are usually not considered in the calculations.

Weaknesses and weak points are usually related to cracks in doors and windows, poor sealed joints, passes for electric and sanitary channeling, construction defects, and interstices. The greater the area of the imperfections, the more they will weaken the acoustic insulation of the wall.

Side or flank transmissions can be as or more important than direct transmission through a wall. They can lead to a significant decrease in the expected insulation. Sound transmission by flanks occurs when the lateral divisors are considerably lighter and/or rigid than the main wall. To solve these issues, the acoustic quality of the laterals needs to be improved to avoiding a poor performance of the solution.

4. Acoustic diffusion

The diffusion of sound is a consequence of the multiple reflections and diffractions that it suffers on irregular surfaces or obstacles to propagation. The diffusers are used to achieve a homogeneous sound field by scattering the reflected acoustic energy in all directions. When a diffuse field is achieved, the acoustic energy is homogeneously and isotropic distributed both in space and in time.

The diffusers allow to correct the early and late reflections and the normal modes of a room.

Unlike the systems of isolation and absorption, in which the most important are the characteristics of the material, the diffusers can be built in any material provided a proper surface design. When a good spatial distribution is achieved, a good temporary dispersion is usually achieved as well.

The design of the surface irregularities can be computed according to different numerical sequences with basis on the main principles of acoustic wave interferences [11].

4.1 Phase diffusers (Schroeder's diffusers)

They are usually built as a sequence of thin linear apertures with different depths. The sound waves penetrate the material, experiment many reflections into the apertures, and emerge from them with a different phase, that is, in a different interference pattern. As a consequence, a good acoustic energy scattering is achieved.

The so-called geometric Schroeder's diffusers or RPG diffusers (reflection phase grating) can be modified in order to be used as acoustic absorbers (Schroeder's absorbers) [12]. Caution is needed to avoid obtaining an undesired behavior (sound absorption instead of sound diffusion).

The most frequent design methods are presented below.

1. **MLS diffusers (maximum length sequence)**

They are designed using a periodic number sequence that decides the position of the apertures on the surface of the material. If the width of the irregularities is reduced, the design frequency of the diffuser will be higher, while if the depth will be higher, the frequencies to be corrected will be lower. The width of the openings is $\lambda/2$ and the depth is $\lambda/4$. This kind of diffusers has a good performance only for one frequency octave.

2. **QRD (quadratic residue diffusers)**

The surface pattern of these diffusers can be one or two dimensional. The first ones have linear openings of the same width and different depths. The depths are obtained by a periodic sequence. The two-dimensional QRD have square cavities of different depths.

3. **PRD (primitive root diffusers)**

They have linear openings of different depths, but they do not suit a periodic pattern because of the number sequence used for the design.

4.2 Crystal-structured diffusers

A crystal structure is an atomic or molecular basic pattern that is identically repeated for many times with the same special distribution and the same orientation. The attraction forces are maxima and are responsible for maintaining the crystal structure.

A sound crystal is also a repetitive structure that causes a recursive scattering by diffraction one time and another [11]. The repetitive properties are the density and the sound propagation speed of the materials that compose the crystal. To meet the desired effect, the wavelength to be controlled should be similar to the dimension of the materials. Crystal sound diffusers have different applications; for example, they are used for the top part of sound barriers.

5. Conclusions

The main acoustic phenomena involved in enclosed ambiences have been presented: acoustic absorption, insulation, and diffusion. A physical explanation of each one has been done, and the main points for their practical application have been summarized.

For the improvement of the acoustic quality of a room, working on its internal surface materials is needed to control the sound absorption (avoiding excessive reverberation) and sound diffusion (avoiding preferred sound paths that generate an inhomogeneous sound field inside).

For controlling the acoustic energy exchange of a room with its external environment, working on the acoustic properties of the boundary enclosure is needed, to reach a reasonable independence from outside and inside acoustic ambiences.

Acknowledgements

This chapter is strongly based on the on Notebook 6 (ISBN: 978-9974-0-1540-1) [13] from the series Environmental Acoustics (ISBN: 978-9974-0-1533-3) from the same author. The series comprises nine notebooks. It has been funded by the National Directorate for the Environmental from the Ministry of Housing, Land Ordering and Environment from Uruguay and published as e-books by the Faculty of Engineering from Universidad de la República (Uruguay).

Author details

Alice Elizabeth González
Department of Environmental Engineering, Facultad de Ingeniería, IMFIA, Universidad de la República, Montevideo, Uruguay

*Address all correspondence to: elizabet@fing.edu.uy

References

[1] Miyara F. Control de Ruido. Own electronic edition; 1999

[2] Blanco Álvarez F. Notas de Curso: Tema II.4. Propiedades Acústicas. España: Universidad de Oviedo; 90 pp

[3] Delgado Pérez L. Paneles para absorción acústica con desechos textiles. Diseño Industrial Textil-Moda [Tesis de grado]. Uruguay: Facultad de Arquitectura, Diseño y Urbanismo, Universidad de la República; Agosto 2017

[4] Roncen R, Fellah ZEA, Simon F, Piot E, Fella M, Ogam E, et al. Bayesian inference for the ultrasonic characterization of rigid porous materials using reflected waves by the first interface. Journal of the Acoustical Society of America. 2018;**144**:210. DOI: 10.1121/1.5044423

[5] Zieliński TG. Normalized inverse characterization of sound absorbing rigid porous media. Journal of the Acoustical Society of America. 2015; **137**(6):3232-3243. DOI: 10.1121/1.4919806

[6] Gower AL, Parnell WJ, Abrahams ID. Multiple waves propagate in random particulate materials. arXiv: 1810.10816v1 [physics.class-ph]; 2018

[7] Liarte García S. Procedimientos para la verificación in situ del aislamiento acústico según el DB-HR del Código Técnico de la Edificación. España: Proyecto Fin de Carrera en Ingeniería Técnica Industrial, Universidad Politécnica de Cartagena; Setiembre 2013

[8] Méndez AM, Stornini AJ, Salazar EB, Giuliano HG, Velis AG, Amarilla BC. Acústica Arquitectónica. Argentina: Universidad del Museo Social Argentino; 1994. 238 pp

[9] Flores Pereita P. Manual de acústica, ruidos y vibraciones. Barcelona, España: Ediciones GYC; 1990. 412 pp. ISBN 978-84-87579-00-4

[10] Recuero López M. Acústica Arquitectónica. Soluciones Prácticas. Madrid, España: Editorial Paraninfo S.A.; 1999. ISBN 978-84-28319-55-3

[11] Fuster Fajardo JM. Estudio de las aplicaciones de los difusores de sonido basados en cristales de sonido [Tesis de Master en Ingeniería Acústica]. España: Universidad Politécnica de Valencia, Escuela Politécnica Superior de Gandía; 2013

[12] Jiménez Arranz G. Estudio y Diseño de Sistemas para el Acondicionamiento Acústico. [Tesis de Master en Ingeniería Acústica de la Edificación y Medio Ambiente]. España: Universidad Politécnica de Madrid; 2013

[13] González AE. Acústica Ambiental. Cuaderno 6: Acondicionamiento Acústico. Montevideo, Uruguay: Universidad de la República, Facultad de Ingeniería, DIA–IMFIA, 2017. ISBN: 978-9974-0-1540-1

Chapter 2

Sound Absorbing Resonator Based on the Framed Nanofibrous Membrane

Klara Kalinova

Abstract

The sound absorbing means are based on a resonance membrane formed by a layer of polymeric nanofibers, which is restricted by a frame. The resonance membrane is then, upon impact of sound waves, brought into forced vibrations, whereby the kinetic energy of the membrane is converted into thermal energy by friction of individual nanofibers, by the friction of the membrane with ambient air and possibly with other layers of material arranged in its proximity. Moreover, part of the kinetic energy of the membrane is transmitted to the frame, to which the membrane is securely attached, and other part is converted into thermal energy due to increased friction in its inner structure, which is caused by the fact that the neighboring parts of the membrane, separated at least partially by the frame or its elements, may vibrate with mutually different periods and/or deviations. The frame is formed by a mesh of grid that can be regular in order to obtain uniform properties over the whole area of the sound absorbing material. The size and shape of the mesh affect the sound absorption or more precisely resonance behavior of the means. To obtain desired sound absorbing characteristics, the resonance membrane is connected to the frame with positive, zero, or negative tension.

Keywords: nanofibers, membrane, resonator, sound absorption, frame

1. Introduction

1.1 Membrane for sound absorption

The term oscillating membrane means a thin plate or foil which has a very small bending stiffness and is located at a distance from the fixed wall. The behavior of such a membrane can be compared to the behavior of a body of a certain weight (represented by a membrane) elastically attached to the spring (represented by an air cushion). The space between the membrane and the rear fixed wall is filled with a porous material that dampens the vibrations of air particles in this space and thus the whole system. Typically, the membrane is selected from such a fabric that its flexural stiffness is much smaller than that of an air cushion [1].

Membrane absorbers are used to absorb low frequency sound. In order to increase the sound absorption, the membrane is positioned at a certain distance parallel to the rigid wall, thus creating an air gap between the wall and the membrane. Coates and Kierzkowski [2] in their paper describe the advantages of thin, light membranes that can replace traditional bulky and economically disadvantageous porous absorbers. The work deals with individual parameters such as the size of the air gap; membrane absorber thickness; its density, flexibility, and, in particular, air flow resistance; and its influence on sound absorption coefficient. The study [3] examines in detail the properties of a simple permeable membrane. The effect of membrane parameters such as basis weight and air flow resistance is clarified. The study is based on the theoretical solution presented in [4] for the perpendicular impact of sound waves. The permeable membrane is characterized by basis weight, tension, and air flow resistance R_h, which is a function of the thickness of the membrane. The influence of air flow resistance R_h to the sound absorption coefficient α occurs especially at higher frequencies. For extreme values, sound absorption coefficient α is zero. This is the case with extremely low air flow resistance R_h when all sound energy passes through the membrane. In the case of a very high R_h value, the membrane becomes impermeable, and all the sound energy is reflected. The optimal R_h value varies with the sound frequency and basis weight of the membrane. The influence of the basis weight is evident especially at lower frequencies. For higher sound frequencies, the value of α is almost constant. Thus, the basis weight loses the effect at higher frequencies, and only R_h plays the dominant role.

In the study [5, 6], the complete form of the analytical solution of the membrane sound absorption coefficient was described. A membrane of infinite dimensions lying in a plane parallel to the fixed wall at a certain distance is envisaged. The membrane characterized by the basis weight and the tension is vibrating by the impact of the plane wave below a given angle of incidence. Both side surfaces of the membrane, the source and back sides, as well as the wall surface, are described by a specific acoustic admittance. The sound absorption coefficient expresses the amount of energy absorbed, including energy losses of different types, which may be caused by different mechanisms at different locations in the system. The frequency of the α maxima, which is caused by the resonance of the system, decreases with the growth of the basis weight. The highest peak of α is recorded in a sample basis weight of 2 kg m^{-2}. With the increasing acoustic admittance of the back side of the membrane, sound absorption coefficient increases in frequency range up to 2 kHz; above this frequency, no effect of surface admittance on the back side, α is constant. Therefore, it can be argued that the sound absorption at higher frequencies is affected mainly by the acoustic admittance of the surface of the source side of the membrane.

A double resonant element, an acoustic element composed of two membranes separated by an air gap, was investigated in studies [7, 8]. Four types of two-layered resonant elements were measured, differing in basis weights [7]. The first membrane positioned in the direction of the incident sound wave always had a substantially smaller (approx. 10x) basis weight over the second membrane. The experimental results were in good agreement with the theoretical model. Furthermore, the influence of the thickness of the air gap, the weight of the two membranes, and their air flow resistance R_h were investigated. The study [8] also monitors the effect of airflow resistance of the first membrane on the acoustic behavior system. In this case, however, the first membrane positioned in the direction of the incident sound waves was permeable, characterized by an air flow resistance R_h.

The permeability of the membrane with optimal air flow resistance improves the absorption properties in the high sound frequency band. At low frequencies, the sound absorption coefficient increases with increasing mass of the first permeable membrane, while it decreases with increasing mass of the second solid membrane. Any effect of membrane mass on sound absorption coefficient is not found at frequencies higher than 2 kHz. At low frequencies the characteristics are independent on air cavity between both membranes. At high frequencies, they are similar to those of a permeable membrane with an air back cavity and a rigid back wall.

The mechanical analogy can be compared to linear electric circuit theory [9]. The electric impedance is defined by the ratio of voltage and current. The acoustic impedance is established analogous to electric impedance as a ratio of sound pressure and acoustic volume velocity. In order to obtain the acoustic impedance of the whole system, the acoustic impedance of the individual elements was first calculated. The relationship was then obtained for calculating the sound absorption coefficient. Two types of glass fiber fabric and a microperforated synthetic membrane were measured to confirm this theory. Measurements took place in both the reverberation room and the impedance tube. The results obtained theoretically are in good agreement with the measured values. A honeycomb structure, or a hexagonal structure, can create a lightweight and stable frame of the acoustic membrane element. Such an absorber studied at work [10] is designed for room acoustics as well as for industrial applications.

In many other studies, various modifications of the oscillating membrane clamped in a circular frame are discussed. Determining the exact solution of oscillations of the circular membrane with inhomogeneous density was studied in works [11–15]. The study [16] focuses on the research of the base frequency of the circular membrane with additional star-shaped distortion originating from the outer edge. It is clear from the results that the base frequency increases with an increasing number of evenly distributed breaches and a length of breach. The article [17] describes the results of the research of the base frequency of a circular membrane that is in contact with water. It is known that the base frequencies of the structures in the water are smaller than in the air, due to the increase in the total kinetic energy of the system due to the presence of water.

1.2 Resonance frequency of membrane

1.2.1 Resonance frequency of circular membrane

A thin circular membrane is a formation that results, for example, from the tension of a thin homogeneous elastic film with a constant basis weight of m_{sq} ($\mathrm{kg\ m^{-2}}$) on a rigid circular frame [18]. By this tension caused by the radially acting force F_r, the membrane gains its stiffness. The radially acting tensile force relative to the unit length of the frame circumference is constant in all directions, and then the tension ν ($\mathrm{N\ m^{-1}}$) is given by the formula:

$$\nu = \frac{F_r}{2\pi R}, \tag{1}$$

where F_r (N) is the total tensioning force and R (m) denotes the radius of the membrane (or the radius of the rigid circular support through which the membrane is tensioned). We expect this tension acting in the plane of the membrane to be the same in all places. Apart from circular membranes, other membrane shapes are used in applications, e.g., elliptical, rectangular membranes.

If the membrane deflects from a normal position (e.g., by acoustic pressure), the membrane that originally formed the plane surface is deformed. If the force that causes the deflection ceases to occur, all the membrane points return; their potential energy, which they get by the deflection, changes into kinetic energy of the moving substantial elements, and the membrane vibrates. If the damping is not taken into account, the membrane will vibrate with free undisturbed vibrations.

If the assumption of axially symmetrical vibrations is fulfilled, then the following relation (2) applies, from which it is possible to determine the base membrane frequencies $f_{0,i}$ (Hz) using the constant of the vid $a_{0,i}$ (equal to 2.4048 for $f_{0,1}$, 5.5201 for $f_{0,2}$, 8.6537 for $f_{0,3}$, 11.97915 for $f_{0,4}$):

$$f_{0,i} = \frac{1}{2\pi R} a_{0,i} C_M, \tag{2}$$

where C_M (m s^{-1}) is the velocity of transverse wave propagating on the membrane given by the relationship

$$C_M = \sqrt{\frac{\nu}{m_{sq}}}, \tag{3}$$

where m_{sq} (kg m^{-2}) is the basis weight of the membrane.

To calculate the membrane base frequencies according to formula Eq. (2), it is necessary first to determine the velocity of the transverse wave propagating on the membrane C_M. This is not possible without knowledge of the radially acting tensioning force v that causes the diaphragm tension on the circular frame. The C_M and ν values are not known, so Eq. (2) cannot be applied to the calculation. By adjusting it, however, the necessary relationships can be obtained.

For determining the angular velocity $\omega_{0,i}$ (s^{-1}), the following formula is commonly used:

$$\omega_{0,i} = 2\pi f_{0,i}. \tag{4}$$

By putting it in relation Eq. (2), the equation Eq. (4) can be rewritten as follows:

$$2\pi f_{0,i} = \frac{a_{0,i} C_M}{R}. \tag{5}$$

After the conversion of this relationship, it is possible to express the relation

$$f_{0,i} = \frac{a_{0,i} C_M}{2\pi R} \Rightarrow \frac{f_{0,i}}{a_{0,i}} = \frac{C_M}{2\pi R}. \tag{6}$$

The ratio $\frac{C_M}{2\pi R}$ is constant since the membrane radius and the velocity of the wave propagation by the membrane do not change. Thus, the ratio of frequency and relevant constant of the vid is constant as follows:

$$\frac{f_{0,1}}{a_{0,1}} = \frac{f_{0,2}}{a_{0,2}} = \frac{f_{0,3}}{a_{0,3}} = \frac{f_{0,4}}{a_{0,4}}. \tag{7}$$

From the abovementioned relationships Eqs. (4)–(6), it is obvious that the radius of the membrane R is inversely proportional to the frequency $f_{0,i}$ and with the increasing radius of the membrane, the own frequency falls.

1.2.2 Resonance frequency of rectangular membrane

The membrane is tensioned in the *x-axis* and *y-axis* direction by a tension ν applied per unit length. The rectangular membrane with sides a and b and axes x and y is tensioned in axial direction by forces:

$$F_x = b.\nu, F_y = a.\nu \tag{8}$$

For the rectangular membrane, its resonant frequency $f_{m,n}$ (Hz) according to the work [18] is given by the relation

$$f_{m,n} = \frac{1}{2} C_M \sqrt{\left(\frac{m}{a}\right)^2 + \left(\frac{n}{b}\right)^2}, \tag{9}$$

where m and n are vids in each axis and a and b (m) are the dimensions of the sides of the rectangle.

The nodal lines of the circle in the simplest case of symmetry are concentric circles, the node lines of the rectangle pointing in the simplest case in the direction of the membrane stresses (perpendicular to the sides of the shape) and dividing the rectangle into the same parts in either direction or in both directions. In a more complex case, the node line is guided along the diagonal rectangle. The constant tension of the membrane ν was achieved by observing the constant conditions during the electrospinning of nanofibrous membrane to the grid support. The optical method according to study [19] determined the base resonant frequency of the circular membrane $f_{0,i}$. Assuming a constant value ν and thus C_M, then the relation Eq. (9) can be modified by assigning the relation Eq. (2) as follows:

$$f_{m,n} = \frac{\pi R f_{0,i} \sqrt{\left(\frac{m}{a}\right)^2 + \left(\frac{n}{b}\right)^2}}{a_{0,i}}. \tag{10}$$

2. Acoustic element design

2.1 Principle of acoustic element

The acoustic element is based on a rigid frame in the form of a perforated plate or a flexible frame in the form of linear shapes or grids, the back side of which covers a thin carrier layer with a nanofibrous membrane which is covered with frames to some extent against mechanical damage. The frame also has a visual function. The element arrangement based on a perforated panel with a nanofibrous layer wherein the area of the nanofibrous membrane is determined by the size and shape of the perforation which, in general, does not necessarily have to be repeated throughout its shape and size, and the element thus consists of many different sheets that allow vibration of the membrane resulting in the unique properties of each vibrating area. The properties of the cavity resonator also enter the system, where the thickness of the plate and its distance from the reflecting surface (wall/ceiling application) are also important in addition to the size and spacing of the hole.

The frequency of the perforated panel f_H (Hz), based on the Helmholtz resonator principle, is according to the studies [1, 20] given by an expression:

$$f_H = \frac{c}{2\pi} \sqrt{\frac{S_D}{S_R l d}}, \tag{11}$$

where c (m s^{-1}) is the sound propagation velocity through the medium (air), S_D (m^2) is the cross-sectional area of the cavity, S_R (m^2) is the area of the resonator (hole spacing), l (m) is the thickness of the perforated plate, and d (m) is the distance from the reflective wall.

Figure 1 illustrates an arrangement of a frame-based element in the form of linear structures (wire construction) overlapping the nanofibrous membrane over its entire back surface. Each shape of the frame borders the area of the oscillating

Figure 1.
The principle of designing the final solution of a frame-based acoustic element in the form of linear formations with a nanofibrous layer in view and cut. The gray color in cross section shows the frame (wire construction), the blue one is nanofibrous resonant membrane, and the red one is adhesive.

membrane, and, in general, the individual frame structure does not need to be repeated over the whole surface, and the element thus consists of many different borders that allow vibration of the membrane resulting in the unique properties of each oscillating surface.

2.2 Technical solution

The acoustic element in this case is placed onto a frame—in the form of a regular grid of glass tapes of a negligible thickness—the back side which is covered by a thin carrier layer with a nanofibrous membrane. The design is based on the solution of a general element, placed onto a frame in the form of linear structures with a nanofibrous layer (**Figure 1**), where the grid provides regular open areas with a given size and shape of the holes and their spacing. **Figure 2** compares the sound absorbing properties of an acoustic element with and without a nanofibrous membrane. The proposed element is then compared with commonly used absorbent material with the best sound absorption results that have been measured (**Figure 3**). From this comparison, it is clear that the developed acoustic element can compete with the material that is available on the market with the best results, even at a lower composite element thickness (also assumed with a possible air gap). Compared nanofibrous composite thicknesses are 30, 40, 50, and 60 mm of foam material. The benefits of the proposed technology are the space between the acoustic element of thickness 1–5 mm and the wall/ceiling, which can be used to install lighting, speakers, etc.

2.3 Experimental

2.3.1 Grid design

The basis for the production of a mesh with different mesh size was the R117 A01 structural reinforcing fabric, manufactured by Saint-Gobain Adfors and

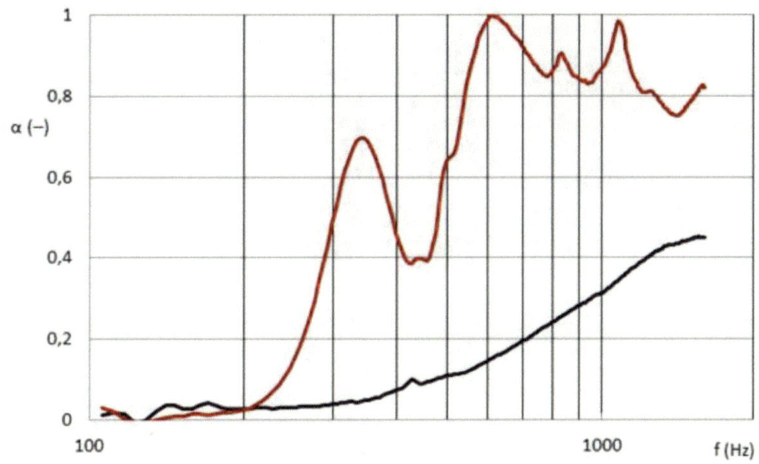

Figure 2.
Frequency dependence of the sound absorption coefficient; comparison of nanofibrous membrane of 0.2 gsm on a carrier of 25 gsm covering the grid of 4 × 4.5 mm mesh size with a thickness of 1 mm with 50 mm air gap (red) and single carrier of 25 gsm covering the same grid of 4 × 4.5 mm mesh size with a thickness of 1 mm with 50 mm air gap (black). Took from [21].

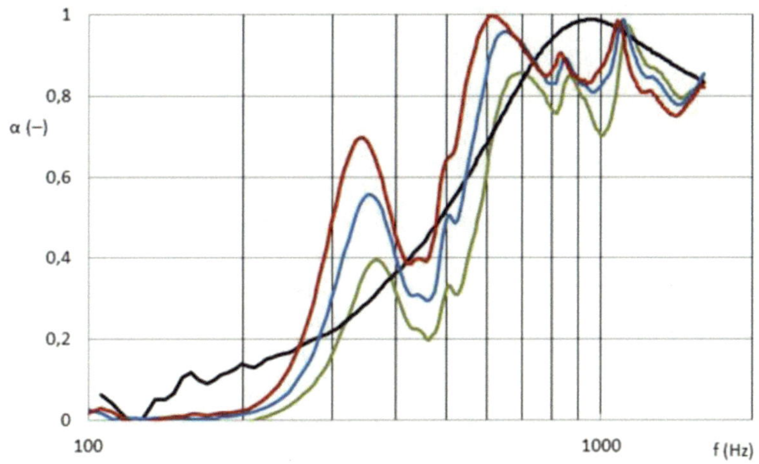

Figure 3.
Frequency dependence of the sound absorption coefficient; comparison of nanofibrous membrane of 0.2 gsm on a carrier of 25 gsm covering the grid of 4 × 4.5 mm mesh size with a thickness of 1 mm with different air gaps (30 mm, green; 40 mm, blue; 50 mm, red) and FOAM 60—foam rectangles with a thickness of 60 mm; width of the base of the rectangle 50 mm foot of the rectangle 40 mm, and top rectangle 60 mm (black). Took from [21].

designed for acrylic plasters. This white grid fabric has a basis weight of 145 g m^{-2} according to the producer, and the mesh spacing has a nominal size of 4.5 × 3.5 mm and a thickness of 0.47 mm.

The smallest mesh size was given by the original dimensions of the grid. The smallest mesh size was determined at 4.1 × 4.3 mm (±0.11 × 0.09 mm) and denoted 1G. The nearest bigger mesh denoted 2G was formed by cutting the weft yarn between two basic meshes at 9.4 × 4.1 mm (±0.11 × 0.11 mm). Another bigger mesh denoted 3G was created by cutting the weft and warp threads between the four basic meshes at 9.0 × 9.4 mm (±0.12 × 0.11 mm). The largest mesh denoted 4G was then formed by cutting out two weft and one warp yarn between the six basic meshes to a size of 9.0 × 14.2 mm (± 0.12 × 0.05 mm) (see **Figure 4**).

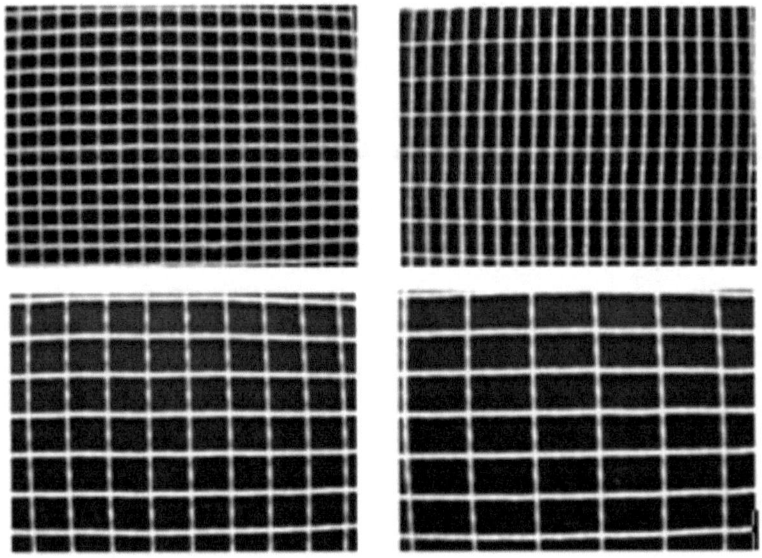

Figure 4.
Photo of applied grids. Rectangle with different side dimensions: 4.1 × 4.3 mm; 9.4 × 4.1 mm; 9.0 × 9.4 mm; 9.0 × 14.2 mm.

2.3.2 Nanofibrous membrane spinning

A nanofibrous layer formed by electrostatic spinning of the polymer from a solution of a 16% polyvinyl alcohol with 40% glyoxal, 85% phosphoric acid, and distilled water was used to produce the membrane. For production of PA6 nanofibrous membrane, a needleless electrospinning method from a cord was employed (Nanospider^{TM}, NS 1WS500U). In this method [22], there is a solution carriage-feeding liquid polymeric material around a moving stainless steel wire. The wire electrode is connected to high-voltage supplier, and on the top, there is a grounded counter electrode. When the applied voltage exceeds a critical value, Taylor cones are then created on the wire surface, oriented toward the counter electrode. PA6 solution jets move toward the collector, and as the solvent evaporates, the PA6 nanofibrous layer is collected on a moving substrate.

The basis weight of the nanofibrous layer is given by the takeoff speed of the backing strip in the electrostatic spinning process. Four basis weights of the nanofibrous layer, namely, 6 g m^{-2} (exact value 5.7 ± 0.2 g m^{-2}), were formed at a corresponding process rate of 0.04 m min^{-1}, 3 g m^{-2} (exact value 2.6 ± 0.15 g m^{-2}) at a corresponding process rate of 0.09 m min^{-1}, 2 g m^{-2} (exact value 2.2 ± 0.03 g m^{-2}) at a corresponding process rate of 0.14 m min^{-1}, and 1 g m^{-2} (exact value 1.7 ± 0.01 g m^{-2}) at a corresponding process rate of 0.18 m min^{-1}. During the spinning on the Nanospider laboratory, a constant temperature of 22.4° C and a relative humidity of 44% RH were maintained with the built-in air conditioner. The spinning process was 0.33–0.34 mA and 50 kV. The distance of electrodes was 150 mm. In **Figures 5** and **6**, the thickness of nanofiber layer of different basis weights was determined using scanning electron microscopy (SEM). A small section of the fiber mat was placed on the SEM sample holder and sputter coated with gold (Quorum Q150R rotary-pumped sputter coater). Carl Zeiss Ultra Plus Field Emission SEM using an accelerating voltage of 1.48 kV was employed to take the SEM photographs.

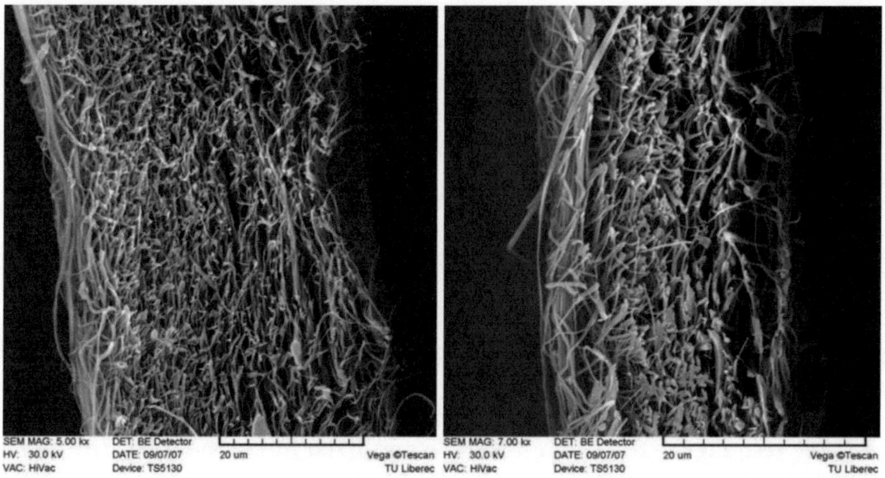

Figure 5.
SEM images of PVA nanofibers of basis weights 6 (5.7 ± 0.2) g m^{-2} (on the left) and 3 (2.6 ± 0.15) g m^{-2} (on the right).

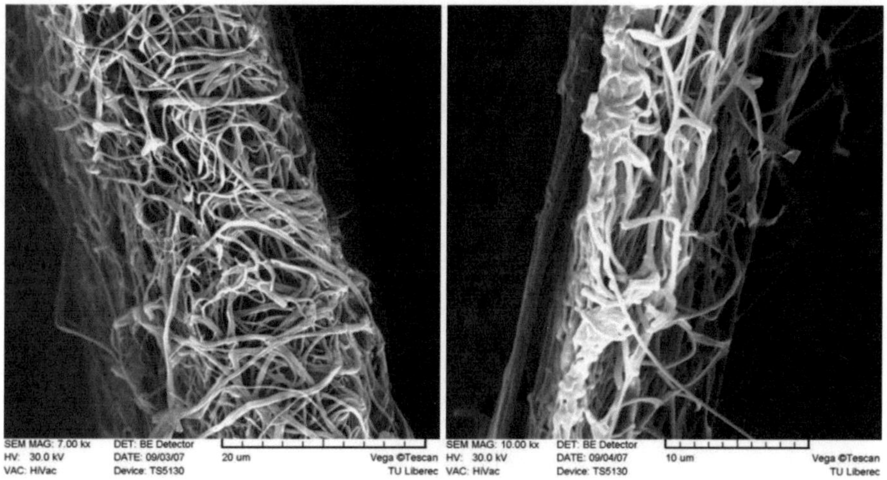

Figure 6.
SEM images of PVA nanofibers of basis weights 2 (2.2 ± 0.03) g m^{-2} (on the left) and 1 (1.7 ± 0.01) g m^{-2} (on the right).

2.3.3 Sound absorption measurement

Two-microphone impedance measurement tube typ. 4206 was used to measure the absorption coefficient in the frequency ranges 50 Hz to 6.4 kHz (standard large tube setup for samples diameter 100 mm: 50 Hz to 16 kHz; standard small tube setup for samples diameter 29 mm: 500 Hz to 6.4 kHz). The test was made according to standard ISO 10534-2. The analyzer (Aubion X.8) generates a random signal which is then amplified by a power amplifier (B&K Typ. 2670, Crown D-75A); frequency is weighted by the frequency weighting unit in the large tube and then applied to the loudspeaker. The analyzer finally measures the response of the two microphones (B&K Typ. 4187) and calculates the frequency response function between these two microphone channels, so the data can be obtained from it. The amount of sound energy which is absorbed is described as the ratio of sound energy absorbed to the sound energy incident and is termed the sound absorption coefficient α. The average of the five measurements was shown. The nanofiber layer was set at a distance of 30 mm from the reflective wall so that the nanofibrous membrane was able to vibrate under the incident sound wave as it is demonstrated on **Figure** 7.

2.3.4 Optical method for determination of resonant frequency

The optical method for determining the resonance frequencies of the membranes [19] was used to calculate the velocity of transverse wave on the homogeneous circular membrane C_M. The main components of the system were the digital camera (Olympus—System i-SPEED2), a LCD display panel of 8.4″, and a transparent tube (see **Figure 8**). The test sample was fixed inside the tube. The incident plane sinusoidal sound wave was excited by a speaker located at end of the tube. The membrane began to oscillate after the impact sound waves reached, and its movement was picked by the high-speed digital camera and in turn was displayed on the LCD.

In order to determine the resonant frequency of the membrane, the 1-1500 Hz frequency range was studied by taking measurements at every 20 Hz to obtain a rough estimate of the resonant frequency. The deflection size of the nanofibrous membrane under the frequency range of 1–1500 Hz was measured using the closed tube. The resonant frequencies of the circular membrane of radius 0.05 m and 1 g m^{-2} basis weight have been detected as can be seen in **Table 1**. The velocity of transverse wave on the homogeneous circular membrane C_M has been determined by the relationship Eq. (2) based on the measured first resonant frequency $f_{0,1}$ equal to 90 Hz.

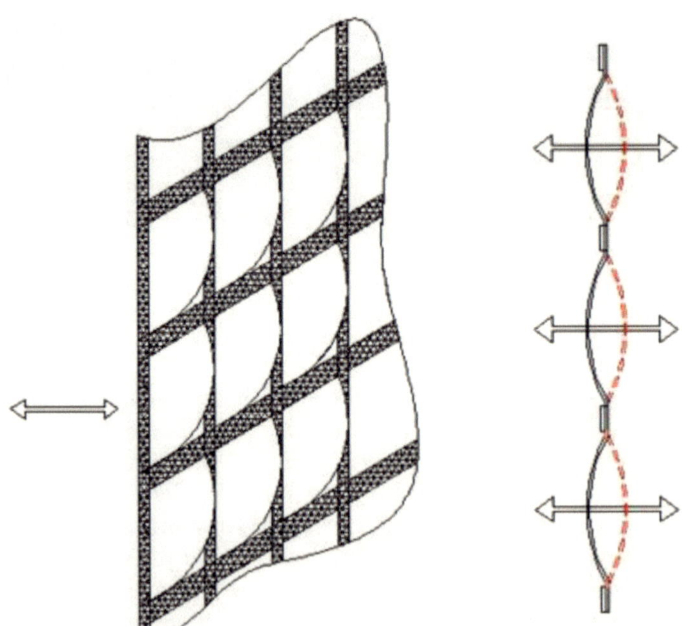

Figure 7.
Illustration of the deflection of vibrating membrane inside the meshes of a grid.

Figure 8.
Scheme of the measuring system. Took from [19].

Circular membrane		$f_{0,i}$ (Hz)	R (m)	C_M (m s^{-1})
$a_{0,1}$	2.4048	90	0.05	11.76
$a_{0,2}$	5 5201	207	0.05	11.76
$a_{0,3}$	8 6537	324	0.05	11.76
$a_{0,4}$	11 7915	441	0.05	11.76

Table 1.
Calculation of the velocity of transverse wave on the membrane C_M based on measured resonant frequency of vibrating circular membrane of radius 0.05 m and 1 g m^{-2} basis weight (a).

3. Sound absorption results

In this arrangement, the idea is connecting the membrane (nanofibrous) resonator together with the mesh frame. The resonant frequencies $f_{m,n}$ of rectangle membrane with a variation of the side dimension are determined according to formula Eq. (10). The calculated resonant frequencies relating to vids m and n are shown in **Tables 2–5**. The dependence of the measured sound absorption coefficient on the sound frequency is shown in the following **Figures 9–16**. In **Figures 9–12**, the measured frequency dependence of the sound absorption coefficient for samples of the same basis weight and different mesh size is compared.

From the curves on **Figures 9–12** describing the frequency dependence of the sound absorption coefficient for samples of the same basis weight and different mesh size, it can be seen that the two clear sound absorption peaks occur in the case of small mesh size (1G and 2G), while in the case of large meshes (3G and 4G), only one clear sound absorption peak exists. It can be seen, therefore, that the mesh size of the grid has a major impact on the amount of the sound absorption coefficient. It can also be observed from **Figure 9** that the nanofibrous membrane of highest basis weight (6 g m^{-2}) applied on the smaller meshes is better at dampening frequencies in the range of approximately 1500 Hz and 4000–5000 Hz, while the nanofiber layer deposited on the larger meshes is better at dampening frequencies of 2500–3000 Hz. Appearance of the peak for the smaller mesh (1G, 2G) at similar frequencies can be explained by the fact that the both meshes have the same width

1G Mesh size 4.1;4.3				*a* (m)	*b* (m)	C_M (m s^{-1})	$f_{m,n}$ (Hz)
m	1	n	1	0.0041	0.0043	11.76	1981
m	1	n	2	0.0041	0.0043	11.76	3087
m	2	n	1	0.0041	0.0043	11.76	3177
m	2	n	2	0.0041	0.0043	11.76	3962
m	3	n	1	0.0041	0.0043	11.76	4514
m	3	n	2	0.0041	0.0043	11.76	5097
m	3	n	3	0.0041	0.0043	11.76	5943
m	1	n	3	0.0041	0.0043	11.76	4345
m	2	n	3	0.0041	0.0043	11.76	5005

Table 2.
Calculation of resonant frequencies of a rectangle membrane with 1 g m^{-2} basis weight of 4.1 × 4.3 mm side dimension; the vids m and n and rectangle dimensions a and b are mentioned.

2G Mesh size 9.4;4.1				*a* (m)	*b* (m)	C_M (m s^{-1})	$f_{m,n}$ (Hz)
m	1	n	1	0.0094	0.0041	11.76	1564
m	1	n	2	0.0094	0.0041	11.76	2935
m	2	n	1	0.0094	0.0041	11.76	1903
m	2	n	2	0.0094	0.0041	11.76	3129
m	3	n	1	0.0094	0.0041	11.76	2361
m	3	n	2	0.0094	0.0041	11.76	3427
m	3	n	3	0.0094	0.0041	11.76	4693
m	1	n	3	0.0094	0.0041	11.76	4347
m	2	n	3	0.0094	0.0041	11.76	4480

Table 3.
Calculation of resonant frequencies of a rectangle membrane with 1 g m^{-2} basis weight of 9.4 × 4.1 mm side dimension; the vids m and n and rectangle dimensions a and b are mentioned.

3G Mesh size 9.0; 9.4				*a* (m)	*b* (m)	C_M (m s^{-1})	$f_{m,n}$ (Hz)
m	1	n	1	0.009	0.0094	11.76	904
m	1	n	2	0.009	0.0094	11.76	1411
m	2	n	1	0.009	0.0094	11.76	1448
m	2	n	2	0.009	0.0094	11.76	1809
m	3	n	1	0.009	0.0094	11.76	2057
m	3	n	2	0.009	0.0094	11.76	2325
m	3	n	3	0.009	0.0094	11.76	2713
m	1	n	3	0.009	0.0094	11.76	1987
m	2	n	3	0.009	0.0094	11.76	2286

Table 4.
Calculation of resonant frequencies of a rectangle membrane with 1 g m^{-2} basis weight of 9.0 × 9.4 mm side dimension; the vids m and n and rectangle dimensions a and b are mentioned.

4G Mesh size 9.0;14.2				*a* (m)	*b* (m)	C_M (m s^{-1})	$f_{m,n}$ (Hz)
m	1	n	1	0.009	0.0142	11.76	773
m	1	n	2	0.009	0.0142	11.76	1055
m	2	n	1	0.009	0.0142	11.76	1370
m	2	n	2	0.009	0.0142	11.76	1547
m	3	n	1	0.009	0.0142	11.76	2003
m	3	n	2	0.009	0.0142	11.76	2127
m	3	n	3	0.009	0.0142	11.76	2320
m	1	n	3	0.009	0.0142	11.76	1403
m	2	n	3	0.009	0.0142	11.76	1803

Table 5.
Calculation of resonant frequencies of a rectangle membrane with 1 g m^{-2} basis weight of 9.0 × 14.2 mm side dimension; the vids m and n and rectangle dimensions a and b are mentioned.

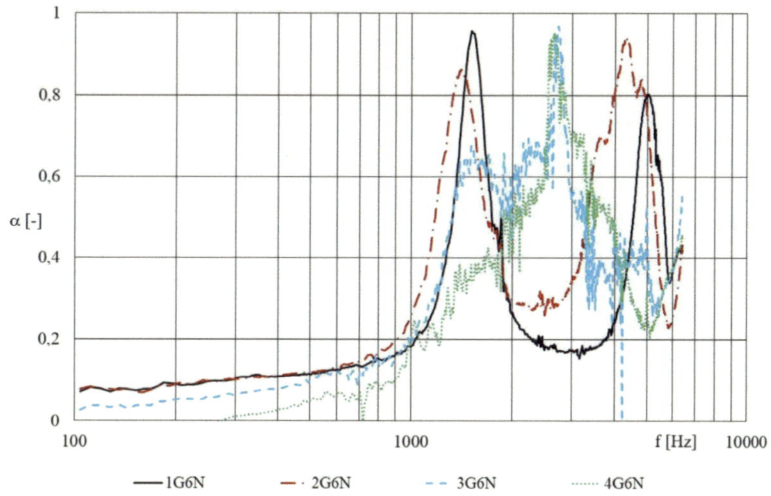

Figure 9.
Frequency dependence of the sound absorption coefficient; comparison of nanofibrous membrane of 6 g m^{-2} covering the grid of different mesh sizes: 1G_4.1 × 4.3 mm; 2G_9.4 × 4.1 mm; 3G_9.0 × 9.4 mm; 4G_9.0 × 14.2 mm. The air gap between the sample of 1 mm thickness and reflective wall was 30 mm.

of 4.1 mm, which is provides the same deviation in the one of axis, regardless of the total area of the mesh. The similar phenomenon can be seen for the highest meshes (3G and 4G) where the constant dimension of meshes is 9 mm with result of identical one of sound absorption peaks for all sample configurations (**Figures 9–12**).

Subsequently, the influence of the basis weight of the nanolayer spun on a grid of the same mesh size onto the maximum values of the sound absorption coefficient in relation to the frequency was examined (see **Figures 13–16**). By checking the grids themselves (gray curve on **Figures 13–16**), almost identical curves of all mesh sizes were found, with only the minimum values of the sound absorption coefficient. It can be said, therefore, that the mass of the carrier grid alone does not significantly influence the course of the sound absorption curves. The basis weight influence on the sound absorption is given by the theory Eqs. (2) and (3) where the resonant frequency of homogenous membrane decreases with its increasing basis weight. The

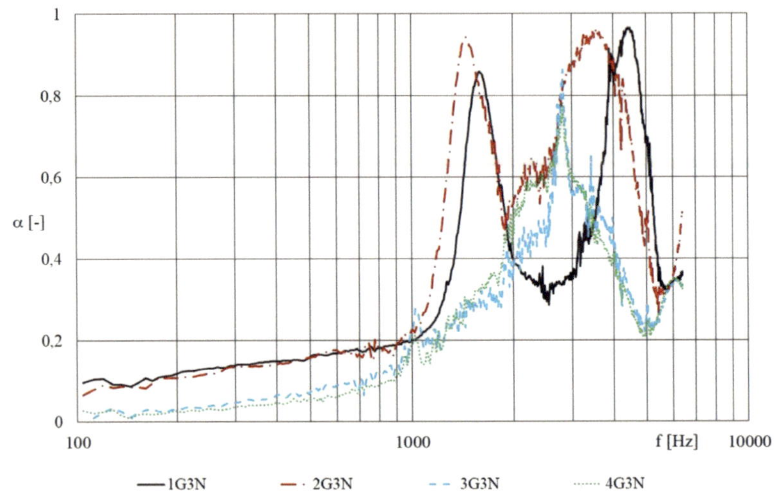

Figure 10.
Frequency dependence of the sound absorption coefficient; comparison of nanofibrous membrane of 3 g m^{-2} *covering the grid of different mesh sizes: 1G_4.1 × 4.3 mm; 2G_9.4 × 4.1 mm; 3G_9.0 × 9.4 mm; 4G_9.0 × 14.2 mm. The air gap between the sample of 1 mm thickness and reflective wall was 30 mm.*

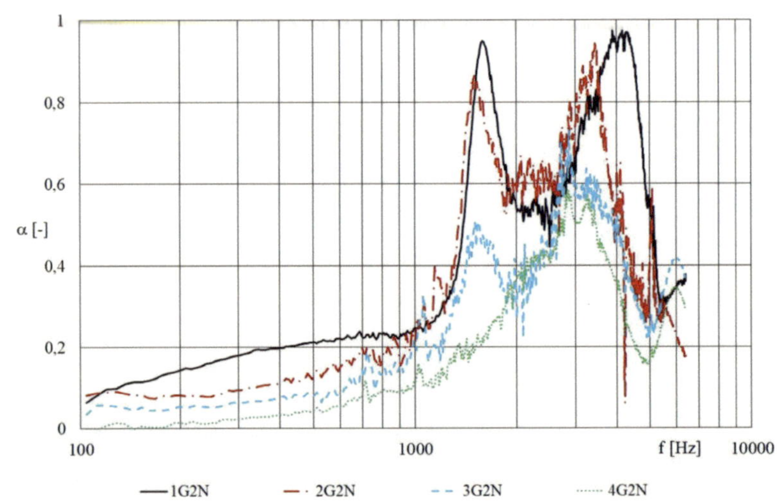

Figure 11.
Frequency dependence of the sound absorption coefficient; comparison of nanofibrous membrane of 2 g m^{-2} *covering the grid of different mesh sizes: 1G_4.1 × 4.3 mm; 2G_9.4 × 4.1 mm; 3G_9.0 × 9.4 mm; 4G_9.0 × 14.2 mm. The air gap between the sample of 1 mm thickness and reflective wall was 30 mm.*

influence of basis weight on the sound absorption coefficient is not clear for these measured configurations. For the smaller meshes (1G and 2G), the sound absorption increases with decreasing basis weight of nanofibrous membrane. Then the antiresonance effect of heavy membrane where the acoustic element loses sound absorption ability (approx. 2500 Hz) occurs due to undamped vibrating membrane. For the higher meshes, the antiresonance effect does not occur because of the grid configuration with the 30 mm air gap does not allow sound absorption below 1500 Hz where the first resonant frequency should cause the first sound absorption peak.

The resonant frequencies with relevant sound absorption coefficients are summarized in **Table 6**. The resonant frequency for membrane of 1 g m^{-2} (lightest

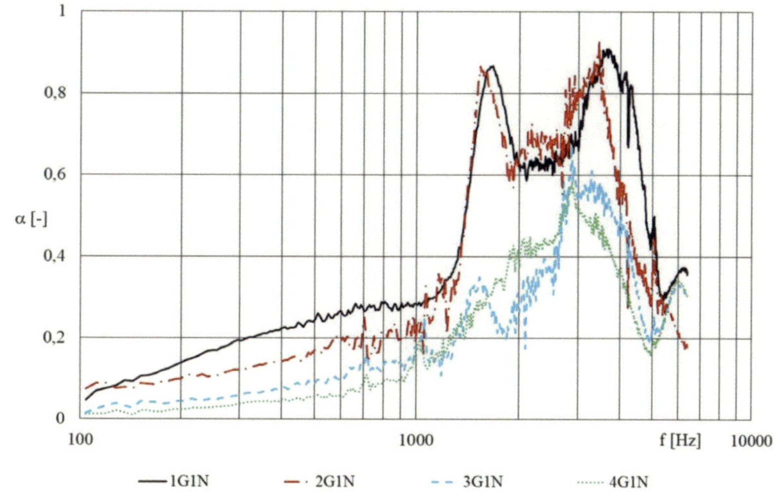

Figure 12.
Frequency dependence of the sound absorption coefficient; comparison of nanofibrous membrane of 1 g m^{-2} covering the grid of different mesh sizes: 1G_4.1 × 4.3 mm; 2G_9.4 × 4.1 mm; 3G_9.0 × 9.4 mm; 4G_9.0 × 14.2 mm. The air gap between the sample of 1 mm thickness and reflective wall was 30 mm.

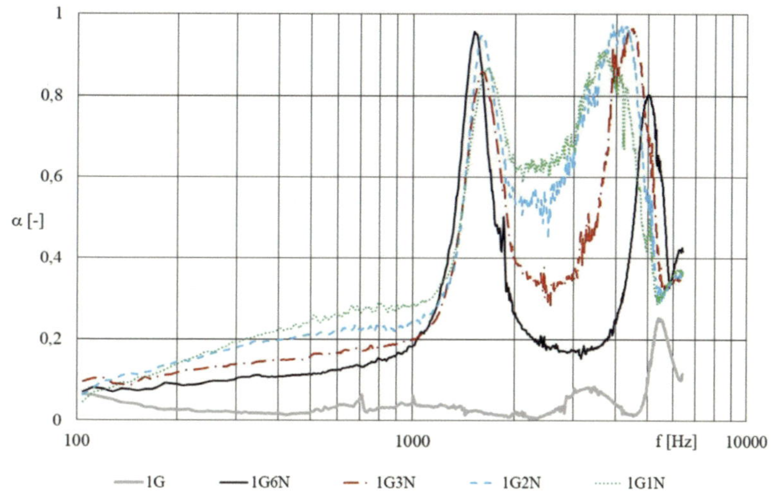

Figure 13.
Frequency dependence of the sound absorption coefficient; comparison of nanofibrous membrane of different basis weights (6N_6 g m^{-2}; 3 N_3 g m^{-2}; 2 N_2 g m^{-2}; 1 N_1 g m^{-2}) covering the grid of mesh size 4 1 × 4.3 mm. 2G_9.4 × 4.1 mm; 3G_9.0 × 9.4 mm; 4G_9.0 × 14.2 mm. The air gap between the sample of 1 mm thickness and reflective wall was 30 mm.

column on the right) can be compared with the calculated values from **Tables 2–5**. The measured values of resonant frequency are in a good agreement for samples with mesh grid 1G where the calculated value of the first and second resonant frequency $f_{1,1}$ = 1981 Hz and $f_{2,2}$ = 3962 Hz (**Table 2**) can be compared with the measured first and second resonant frequencies f_1 = 1672 Hz and f_1 = 3696 Hz (**Table 6**) given by the sound absorption peaks from **Figure 12**. Analogous to the results, the samples with mesh grid 2G where the calculated value of the first and second resonant frequencies $f_{1,1}$ = 1564 Hz and $f_{2,2}$ = 3129 Hz (**Table 2**) can be compared with the measured first and second resonant frequencies f_1 = 1536 Hz and

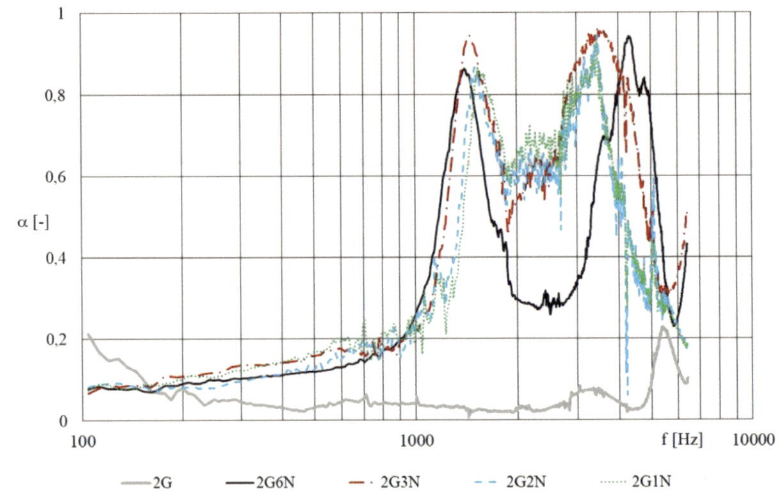

Figure 14.
Frequency dependence of the sound absorption coefficient; comparison of nanofibrous membrane of different basis weights (6N_6 g m^{-2}; 3 N_3 g m^{-2}; 2 N_2 g m^{-2}; 1 N_1 g m^{-2}) covering the grid of mesh size 9.4 × 4.1 mm. The air gap between the sample of 1 mm thickness and reflective wall was 30 mm.

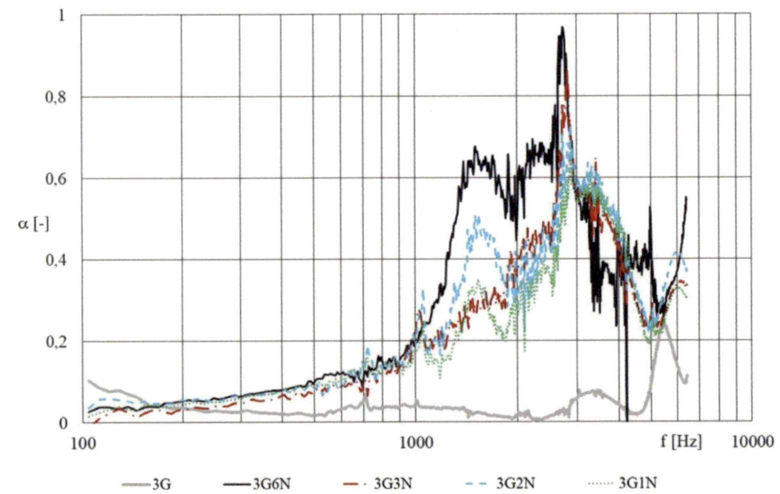

Figure 15.
Frequency dependence of the sound absorption coefficient; comparison of nanofibrous membrane of different basis weights (6N_6 g m^{-2}; 3 N_3 g m^{-2}; 2 N_2 g m^{-2}; 1 N_1 g m^{-2}) covering the grid of mesh size 9.0 × 9.4 mm. The air gap between the sample of 1 mm thickness and reflective wall was 30 mm.

f_1 = 3448 Hz (**Table 6**) given by the sound absorption peaks from **Figure 12**. The samples with mesh grid 3G where the calculated value of first and second resonant frequency $f_{1,1}$ = 904 Hz and $f_{2,2}$ = 1809 Hz (**Table 2**) can be compared with the measured first and second resonant frequencies f_1 = 1056 Hz and f_1 = 1536 Hz (**Table 6**) given by the sound absorption peaks from **Figure 12**. The clear sound absorption peaks of sample with mesh grid 4G do not occur. That is why the comparison cannot be done.

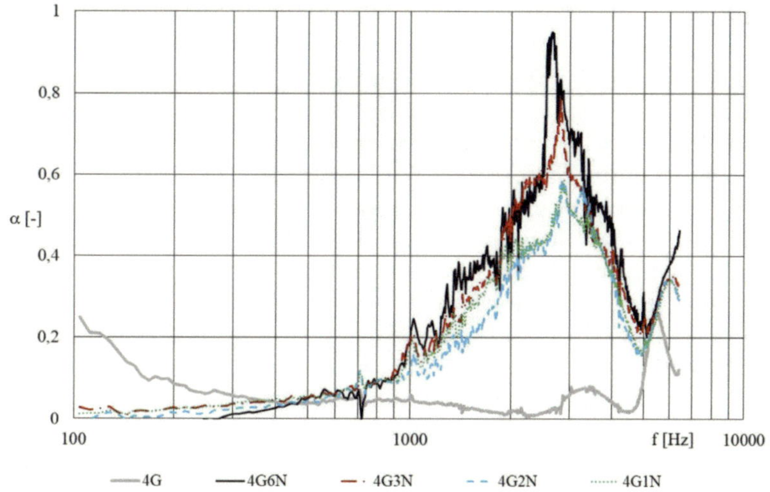

Figure 16.
Frequency dependence of the sound absorption coefficient; comparison of nanofibrous membrane of different basis weights (6N_6 g m^{-2}; 3 N_3 g m^{-2}; 2 N_2 g m^{-2}; 1 N_1 g m^{-2}) covering the grid of mesh size 9.0 × 14.2 mm. The air gap between the sample of 1 mm thickness and reflective wall was 30 mm.

1G mesh with membrane 6 N			**1G mesh with membrane 3 N**			**1G mesh with membrane 2 N**			**1G mesh with membrane 1 N**		
α	f_1 (Hz)	f_2 (Hz)	α	f_1 (Hz)	f_2 (Hz)	α	f_1 (Hz)	f_2 (Hz)	α	f_1 (Hz)	f_2 (Hz)
0.858	1600		0.858	1600		0.95	1592		0.867	1672	
0.801		5008	0.966		4432	0.974		3896	0.906		3696
2G mesh with membrane 6 N			**2G mesh with membrane 3 N**			**2G mesh with membrane 2 N**			**2G mesh with membrane 1 N**		
α	f_1 (Hz)	f_2 (Hz)	α	f_1 (Hz)	f_2 (Hz)	α	f_1 (Hz)	f_2 (Hz)	α	f_1 (Hz)	f_2 (Hz)
0.944	1456		0.943	1456		0.95	1592		0.867	1536	
0.941		4312	0.958		3560	0.943		3448	0.934		3448
3G mesh with membrane 6 N			**3G mesh with membrane 3 N**			**3G mesh with membrane 2 N**			**3G mesh with membrane 1 N**		
α	f_1 (Hz)	f_2 (Hz)	α	f_1 (Hz)	f_2 (Hz)	α	f_1 (Hz)	f_2 (Hz)	α	f_1 (Hz)	f_2 (Hz)
0.677	1512		0.863	2816		0.51	1536		0.25	1056	
0.968		2728	0.651		3424	0.731		2872	0.349		1536
4G mesh with membrane 6 N			**4G mesh with membrane 3 N**			**4G mesh with membrane 2 N**			**4G mesh with membrane 1 N**		
α	f_1 (Hz)	f_2 (Hz)	α	f_1 (Hz)	f_2 (Hz)	α	f_1 (Hz)	f_2 (Hz)	α	f_1 (Hz)	f_2 (Hz)
0.95	2656		0.783	2816		0.586	2840		—	—	
—		—	—		—	0.57		3272	—		—

Table 6.
Measured resonant frequencies of a rectangle membrane of different basis weights and different side dimensions. The sound absorption coefficient α is mentioned for each resonant frequency rectangle membrane.

4. Conclusions

This chapter has studied the behavior of nanofibrous membrane in a synergy with a grid over the sound waves incidence. The sound absorption coefficient of a frame-based element in the form of linear structures overlapping the nanofibrous membrane over its entire back surface has been measured. A nanofibrous membrane having a basis weight of 6, 3, 2, and 1 g m^{-2} covering a rectangle grid with the different side dimensions, 4.1 × 4.3 mm, 9.4 × 4.1 mm, 9.0 × 9.4 mm, and 9.0 × 14.2 mm, has been placed at a constant distance of 30 mm from the reflective wall. Two-microphone impedance measurement tube typ. 4206 was used to measure the absorption coefficient in the frequency ranges 50 Hz to 6.4 kHz according to standard ISO 10534-2.

By measuring the sound absorption of nanofibrous layers with the same area weights covering the grids of different mesh sizes, it was found that the nanofiber layer applied on smaller mesh achieves better sound absorption where two absorption peaks occur at approximately 1500 and 3500 Hz, while nanofibrous membrane applied to the larger mesh grids does not absorb the sound of these frequencies enough. From the analysis of sound absorption measurement results, it can be assumed that the mass of the carrier grid alone does not significantly influence the course of the sound absorption curves. The influence of basis weight on the sound absorption coefficient is not clear for all measured configurations. For the smaller meshes, the sound absorption increases with decreasing basis weight of nanofibrous membrane. Then the antiresonance effect of heavy membrane where the acoustic element loses sound absorption ability (approx. 2500 Hz) occurs due to undamped vibrating membrane. The maximum values of the sound absorption coefficient occur at resonant frequencies of the sound absorption system. The resonant frequencies of 1 g m^{-2} membrane have been compared with the calculated values, whereas the calculated and measured values are in a good agreement.

Acknowledgements

The results of this project LO1201 were obtained through the financial support of the Ministry of Education, Youth and Sports in the framework of the targeted support of the "National Programme for Sustainability I" and the Technology Agency of the Czech Republic within the project "Broadband components with resonant nanofibrous membrane for room acoustics" no. TH02020524.

Conflict of interest

The author(s) declared no potential conflicts of interest with respect to the research, authorship, and/or publication of this article.

Author details

Klara Kalinova
Technical University of Liberec, Liberec, Czech Republic

*Address all correspondence to: klara.kalinova@tul.cz

References

[1] Kolmer F, Kyncl J. Prostorová akustika. 1st ed. Prague: SNTL; 1980. ISBN 04-514-80

[2] Coates M, Kierzkowski M. Acoustic textiles - lighter, thinner and more sound-absorbent. Technical Textiles International. 2002;**11**(7):15-18

[3] Sakagami K et al. Detailed analysis of the acoustic properties of a permeable membrane. Applied Acoustics. 1998; **54**(2):93-111

[4] Takahashi D, Sakagami K, Morimoto M. Acoustic properties of permeable membranes. The Journal of the Acoustical Society of America. 1996;**99**: 3003-3009

[5] Sakagami K et al. Sound absorption of a cavity-backed membrane : A step towards design method for membrane-type absorbers. Applied Acoustics. 1996; **49**(3):237-247

[6] Sakagami K et al. Acoustic properties of an infinite elastic plate with a back cavity. Acustica. 1993;**78**:288-295

[7] Kiyama M et al. A basic study on acoustic properties of double-leaf membranes. Applied Acoustics. 1998; **54**(3):239-254

[8] Sakagami K, Kiyama M, Morimoto M. Acoustic properties of double-leaf membranes with a permeable leaf on sound incidence side. Applied Acoustics. 2002;**63**(8):911-926

[9] Kang J, Fuchs HV. Predicting the absorption of open weave textiles and micro-perforated membranes backed by an air space. Journal of Sound and Vibration. 1999;**220**(5):905-920

[10] Ackermann U, Fuchs HV, Rambausek N. Sound absorbers of a novel membrane construction. Applied Acoustics. 1988;**25**(3):197-215

[11] Wang CY. Some exact solution of the vibration of non-homogenous membranes. Journal of Sound and Vibration. 1998;**210**(4):555-558

[12] Gottlieb HPW. Exact solutions for vibrations of some annular membranes with inhomogeneous radial densities. Journal of Sound and Vibration. 2000; **233**(1):165-170

[13] Jabareen M, Eisenberger M. Free vibrations of non-homogenous cirkular and annular membranes. Journal of Sound and Vibration. 2001;**240**(3): 409-429

[14] Wang CY. Vibration of an annular membrane attached to a free rigid core. Journal of Sound and Vibration. 2003; **260**(4):776-782

[15] Pinto F. Analytical and experimental investigation on a vibrating annular membrane attached to a central free, rigid core. Journal of Sound and Vibration. 2006;**291**(3):1278-1287

[16] Wang CY. Fundamental models of a cirkular membrane with radial constraints on the boundary. Journal of Sound and Vibration. 1999;**220**(3): 559-563

[17] Kwak MK. Vibration of circular membranes in contact with water. Journal of Sound and Vibration. 1994; **178**(5):688-690

[18] Škvor Z. Akustika a elektroakustika. 1st ed. Academia Praha: Prague; 2001. ISBN 80-200-0461-0

[19] Kalinova K, Ozturk MK, Komarek M. Open and closed tube method for determination of resonance frequencies of nanofibrous membrane. The Journal

of The Textile Institute. 2016;**107**(8): 1068-1078. DOI: 10.1080/ 00405000.2015.1083353

[20] Randeberg RT. Perforated panel absorbers with viscous energy dissipation enhanced by orifice design. PhD [thesis], 1st edition. Trondheim: NTNU; 2000. Available from: https://brage.bibsys.no/xmlui/bitstream/handle/11250/249798/125365_FULLTEXT01.pdf?sequence=1&isAllowed=y

[21] Kalinova K. A sound absorptive element comprising an acoustic resonance Nanofibrous membrane. Recent Patents on Nanotechnology. 2015;**9**(1):61-69. ISSN 1872-2105

[22] Jirsak O, SanetrnikF, et al. Method of nanofibres production from a polymer solution using electrostatic spinning and a device for carrying out the method. Patent WO2005024101 (US2006290031); 2005

Chapter 3

Automated Classification of Microwave Transmitter Failures Using Virtual Sensors

Ayax D. Ramirez, Stephen D. Russell, David W. Brock and Narayan R. Joshi

Abstract

Each year, nearly $100 M is spent replacing high-power microwave tubes in the fleet. In many cases (estimated at over 25%), tubes that are operating perfectly are inadvertently replaced because there are insufficient in-situ monitoring equipment available to diagnose specific problems within the system. High-power microwave vacuum tubes used in radar or communications systems have minimal condition-based maintenance capability and no means to identify specific component failures. This chapter presents the results from a system that uses cathode current and acoustic emission sensors combined as a virtual sensor to locate and classify microwave transmitter failures. Data will be shown which differentiate the failure mode from subsystems on a radar klystron and from a communications system magnetron. The use of the integrated condition assessment system (ICAS) to acquire and track virtual sensor data will also be described. These results offer promise of a low-cost, nonintrusive system to monitor microwave transmitters, which correctly identifies component failures avoiding incorrect replacement of high-value klystrons, magnetrons, or traveling wave tubes. This advanced technique also offers the possibility of developing built-in prognostic capabilities within the radar system to provide advanced warning of a system malfunction.

Keywords: microwave tubes, acoustic emission, virtual sensor, magnetron, klystron, integrated condition assessment system

1. Introduction

Microwave vacuum tubes used in high-power radar and communications systems have a lifetime of a few thousand active hours before refurbishment is required. When one of these microwave vacuum tubes fails, it is generally impossible to determine the sequence of events leading to its failure. Each year, nearly $100 M is spent replacing high-power microwave tubes in the fleet. In many cases (estimated at over 25%), tubes that are operating perfectly are inadvertently replaced because there is insufficient in-situ monitoring equipment available to diagnose specific problems within the system. This results in high maintenance and refurbishing costs.

At present, microprocessor-based systems with sensors are designed to monitor tube performance, provide tube protection, and record a comprehensive tube failure history. A major limitation of these systems results from the small amount of time available during the inter-pulse period of the tube for data buffering and fault analysis. The present monitoring systems work well if the microwave tube is operated with 200 or less pulses per second (pps). Normally, the radar tubes are operated at up to 1000 pps with pulse duration of a microsecond. Increasing the A/D conversion speed will, in some cases, make the situation worse, since it increases the amount of data that must be transferred and analyzed during the small time interval available. These high vacuum devices (10^{-7}–10^{-8} Torr) have electrode voltages that can run up to more than 10 kV, while their heat dissipation ranges from 100 W to 10 kW. The complexity of these systems makes them very expensive to produce, maintain, and replace. This provides a motivation for the development of alternative, more effective monitoring and diagnostic techniques.

In recent years, research has established acoustic emission (AE) sensing as a very effective technique for machine condition monitoring and analysis. This technique has been tested and evaluated in a variety of systems as an alternative to conventional techniques. A novel application of this technique is the in-situ performance monitoring of high-power microwave (HPM) tubes. This report addresses two questions: (1) Would the microwave radar tubes operating under normal or abnormal conditions be able to generate AE signals? (2) If so, can the observed signals provide signatures to discriminate among different types of failures?

Acoustic emission (AE) may be defined as stress or pressure waves generated during dynamic processes in materials. AE is elastic energy that is spontaneously released by materials when they undergo deformation and is typically generated in the form of ultrasound waves created by local mechanical instabilities within the material. AE is generally detected by means of ultrasonic transducers coupled to the material with a suitable coupler to decrease impedance mismatch. Among many mechanisms that produce AE activity, the principal mechanisms are crack initiation and growth, magneto-mechanical realignment or growth of magnetic domains (Barkhausen effect), dislocation movements, twining, phase changes, fracture of brittle inclusions, fiber breakage in composite materials, chemical activity, and cavitation. Some stimuli are necessary to trigger acoustic emissions. Stress, a major type of stimuli, may be mechanically applied, thermally generated, or caused by a changing magnetic field. Acoustic emission could thus act as a passive nondestructive technique (NDT) and be used to monitor and analyze normal and abnormal performance of microwave vacuum tubes.

The research presented in this paper demonstrates the detection of anomalous RF pulses and system failures using acoustic emission and magneto resistive or inductively coupled current sensors. It also demonstrates the ability to discriminate among the different types of failures. This innovative system has been tested on a klystron as part of an AN/SPS-49(V)5 radar system and on a radar system magnetron (Model 2J55). An added feature of this innovative system is the fact that the outputs from the sensors have been successfully interfaced with the ICAS (Integrated Condition Assessment System) software currently used by the U.S. Navy.

Once the output of the sensors was integrated with the ICAS software, a method was developed for integrating a plurality of sensor data in such a way to produce greater information than any individual sensor or combination of sensors. This method is particularly useful for detecting and predicting failures and for life cycle monitoring in microwave vacuum devices. This method has been defined as a virtual sensor.

2. AE as an advanced nondestructive technique

The fundamental principle of this method is based on the phenomenon of the generation of an acoustic pulse when a shock wave is generated inside a solid [1–4].

Generally, AE stress waves disperse throughout the medium until interaction with an interface or boundary produces a reverberating field. Although this energy will be mostly absorbed by the medium, some of the energy can be detected by the use of a high-frequency piezoelectric ceramic transducer. Assuming that all mechanisms of energy loss in the structure and measurement system are constant, the measured electrical signal energy from the transducer is proportional to the AE event energy [5],

$$AE_{energy} \propto \int_0^T V(t)^2 dt \tag{1}$$

where T is the time length of transient signal produced and V(t) is the transient voltage. Thus, the measured electrical signal energy is often referred to as the AE energy descriptor and is written as

$$E_{AE} \propto \left(\frac{1}{R}\right) \int_0^T V(t)^2 dt \tag{2}$$

where E_{AE} refers to the AE energy measured in the transducer and R is the impedance of the complete measuring circuit. The power of the acoustic emission signal of the detected event is proportional to the power of the source event. The advantage of energy measurement over ring down counting is that energy measurements can be directly related to important physical parameters without having to model the acoustic emission signal. Energy measurements also improve the acoustic emission measurement when emission signal amplitudes are low. Squaring the measured energy signal produces a simple pulse from a burst signal and leads to a simplification of event counting.

3. Causes of microwave tube failures

There are several well-known causes of microwave tube failure. These include:

3.1 Cathode emission decrease

A decrease in emission normally results in lowering of both the upper and lower mode boundaries. When this shift downward in current becomes significant, the operating point current has to be adjusted downward to avoid instability and oscillation. The operating point is adjusted periodically during the life of the tube. The tube remains operable until either its output power is low or the shift in mode boundaries due to cathode emission precludes stable operation over the frequency band. Cathode emission degradation is a long-term event that requires regular and careful monitoring.

3.2 Loss of vacuum

Loss of vacuum is a catastrophic failure, which may not be determined with electrical monitors. From experience, it is known that the heater bushing is the predominant failure. Other key vacuum seals include the RF input and output ceramic-to-metal seals.

3.3 Heater failure

Heater failure is a catastrophic event and results in inability to start cathode emission. Shorted heaters are more typical than open heaters. The cause of this fault cannot be determined with electrical monitors.

3.4 Other

Seal failures and inverse pulses are termed as transient events causing tube failures.

4. Application of the AE technique to klystron tubes

The results presented in this section were obtained at the Naval Surface Warfare Center, Port Hueneme Division, Dam Neck Detachment (NSWC PHD DN), and the Fleet Training Center (FLETRACEN) San Diego [6]. A simplified description of the field test unit utilized in these experiments is shown in **Figure 1**.

The radar system utilized during this test was the AN/SPS49 (V)5 radar with a 5-cavity klystron amplifier. The pulse repetition frequency (PRF) was set at 213 pps. This unit typically generates two pulses with durations of 2 and 125 μs, respectively. This klystron unit had 48 channels between 851 and 942 MHz. Channels 1–16 cover the low band; channels 17–32 cover the middle band; and channels 33–48 cover the high band. Channel 8, with a center frequency of 894.33 MHz, was frequently utilized

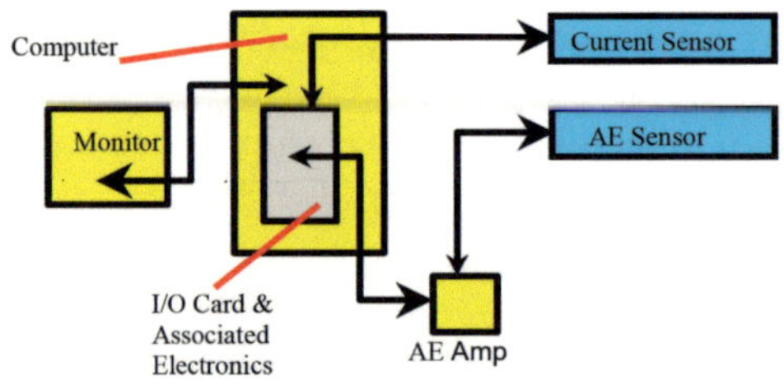

Figure 1.
Simplified field test unit.

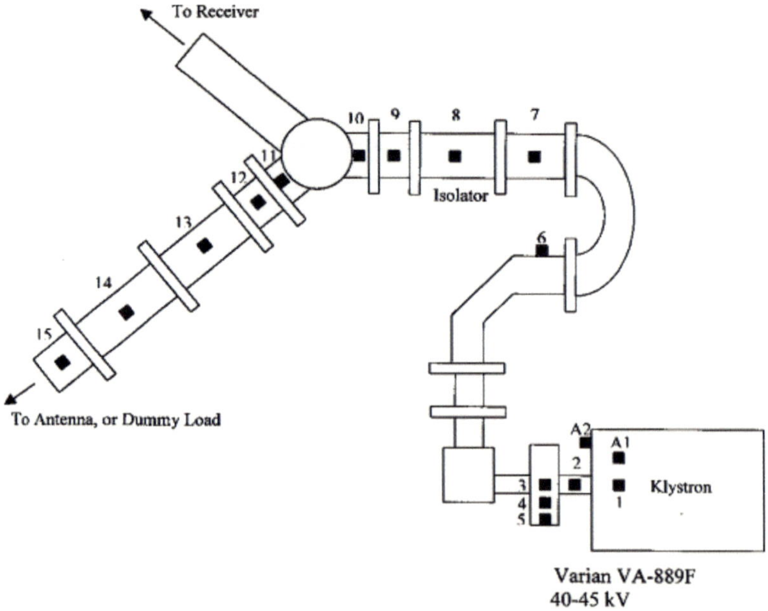

Figure 2.
Sensor location on klystron system.

in this experiment. The klystron unit was connected to a dummy load at the end of the waveguide. The dummy load was cooled by circulating water. During normal operation, the cathode was held at −42 kV with respect to the grounded collector.

The cathode current sensor was connected directly to the control panel of the radar transmitter. The acoustic emission system included a general-purpose R50 transducer, a preamplifier, and a post amplifier, all supplied by Physical Acoustics Corporation. **Figure 2** shows the different locations on the radar, denoted by black squares, where the AE transducer was attached in order to collect the acoustic emissions generated during normal and abnormal stressed operation. Locations include the coax, low pass filter, RF isolator, and several portions of the waveguide. A LeCroy 9354AM 500 MHz digital oscilloscope was used to detect and store signals from the generated acoustic emission, the input (RF drive) of the klystron amplifier, and the beam current sensor.

Figures 3 and **4** show typical signature signals from the sensors for the AE detection of the RF pulse and the cathode current (CC) for different locations on the klystron system under normal operation. These plots show the strong correlation between acoustic emission and RF emission.

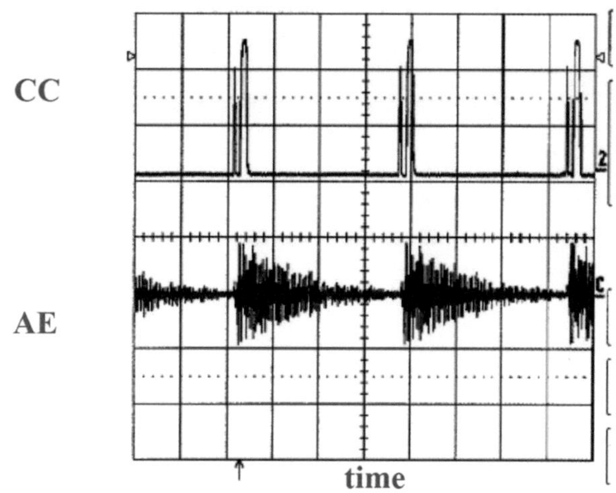

Figure 3.
Signature signals from acoustic and cathode current sensors at 45° waveguide bend, location 6.

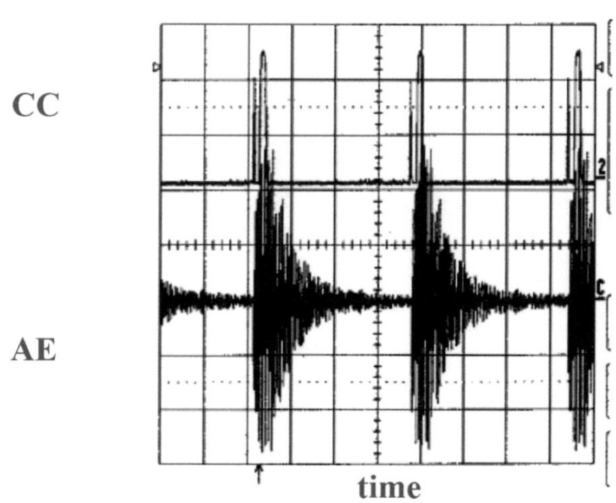

Figure 4.
Signature signals from acoustic and cathode current sensors at RF isolator, location 8.

A threshold amplifier was used to discriminate between normal and anomalous current pulses. Further details of the experiments can be found in Ref. [6].

Anomalous pulses were generated by several methods, including increasing the klystron voltages above normal operating specifications, up to 45 kV. This increase in voltage generated anomalous cathode current and acoustic pulses. **Figure 5** shows a typical anomalous current pulse under stress conditions.

Figure 6 shows a detection of a fault in the system induced by stress conditions. An anomalous cathode current pulse was detected. The corresponding acoustic emission pulse shows a slight increase in the detected acoustic signal. **Figure 7** shows the detection of a very different type of fault. In this case, an anomalous cathode current pulse was detected during a crowbar fault but no acoustic emission was detected. In this case, a catastrophic failure of a modulator tube was responsible for the fault. The absence of acoustic energy along with the detection of an

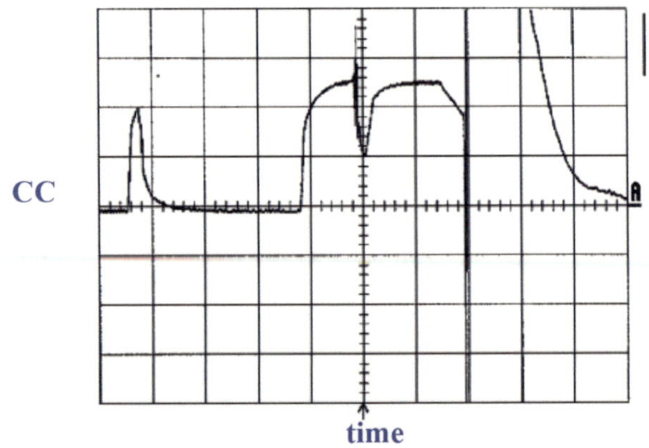

Figure 5.
Anomalous current pulse under stress conditions.

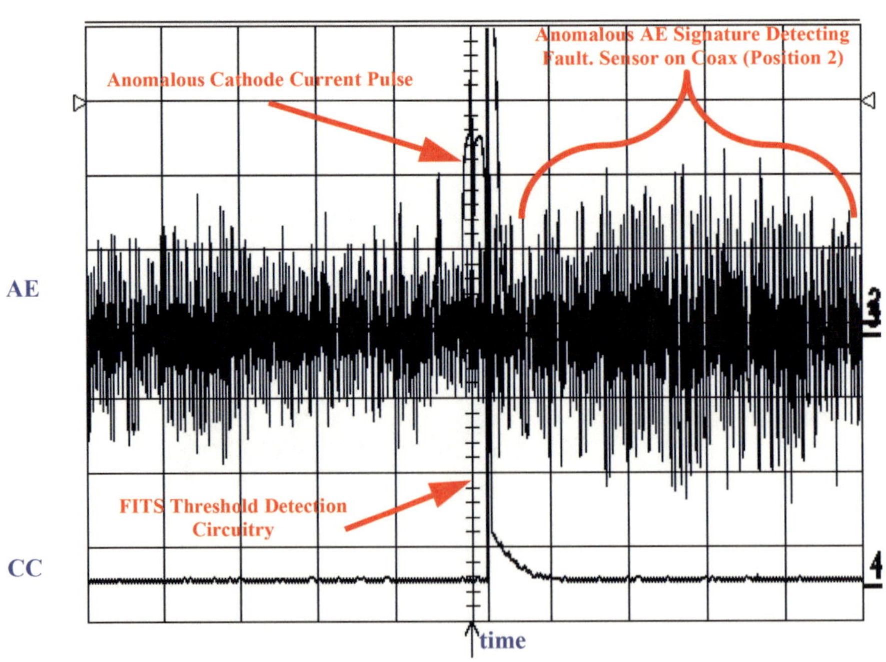

Figure 6.
Fault detection: anomalous current pulse with corresponding change in AE.

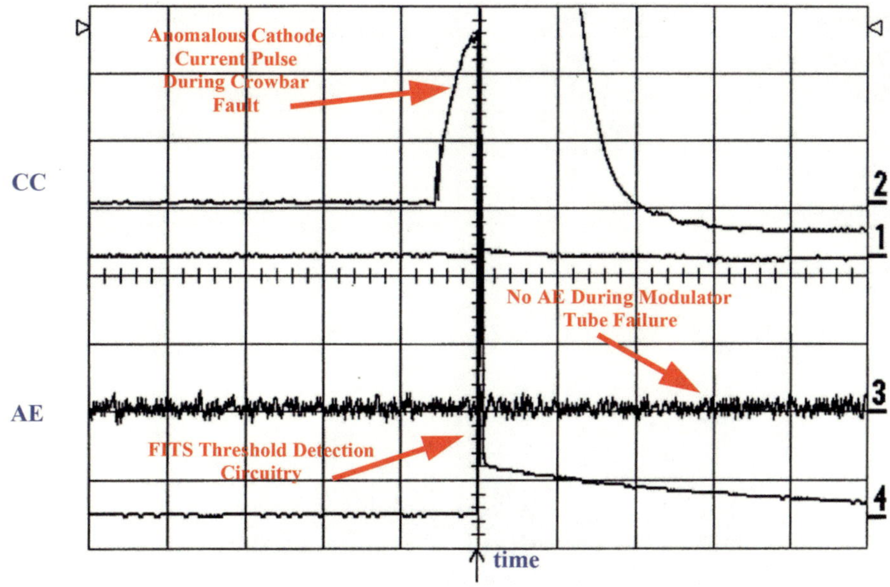

Figure 7.
Fault detection: anomalous current pulse with no AE.

anomalous cathode current pulse provides a possible signature for modulator tube failure. The system demonstrated the ability to discriminate between normal and anomalous pulses in klystron devices using current and acoustic sensors, and it showed a correlation between the two types of detection. The system also demonstrated the ability to characterize different types of failures.

5. Application of the AE technique to magnetron tubes

This section presents the effects on AE during normal and abnormal functioning of a magnetron. The experiment was conducted at the SPAWAR Systems Center, San Diego microwave radar tube laboratory [7]. The proper operation of a magnetron device depends upon the amplitude of the pulsed voltage applied, the temperature, the filament voltage, and the proper loading of the magnetron by the RF system. The normal operating parameters of the magnetron tube (2J55) utilized for this experiment were intentionally changed in order to stress the magnetron tube and induce abnormal functioning. **Table 1** shows the magnetron parameters for the experiments described below.

It was observed that when the tube was stressed, it typically produced an anomalous current pulse with a resulting anomalous RF pulse. The pulses were captured utilizing the masking feature of a 500-MHz digital LeCroy 9354AM oscilloscope, where a mask is defined based on a normal pulse shape, and pulses which lie outside the mask are captured. A fast Fourier transform was performed on the captured AE signal.

Figures 8 and **9** were obtained by using a type S9208 AE transducer, and they show the normal and abnormal pulses, respectively. In both cases, the pulse rate was adjusted to 100 pps, and the oscillator filament primary voltage was set at the normal value of 115 V. Channel 3 on the oscilloscope was connected to the output of the acoustic emission post amplifier and is shown in the upper trace. Channel B was connected to the current sensor of the magnetron, and it shows a zoom trace of the current pulse (see lower curve in figures). The mask was set on channel D

Parameter	Typical values	Stressed condition
Filament voltage	6.3 V AC	3.2 V AC
Pulse rate	1000 pps	1000–10,000 pps
Case temperature	95°F	Cooling fan shut off, temperature increased to over 150°F.
Reflected power	0%	100%
Pulse voltage	15 kV	15–21 kV

Table 1.
Magnetron parameters.

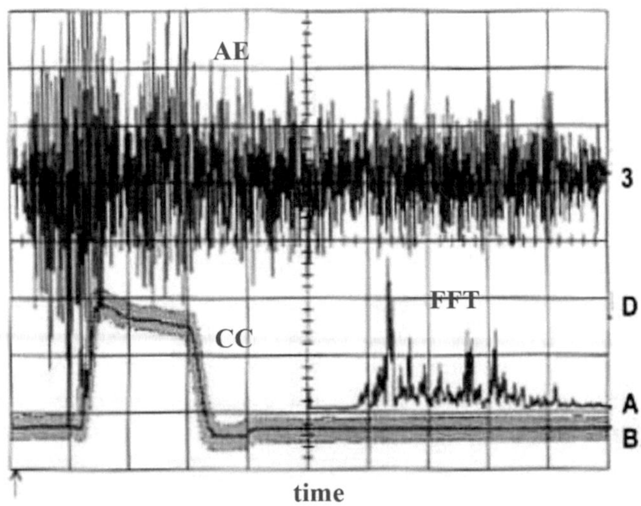

Figure 8.
Magnetron under normal operation with pulse voltage of 15 kV.

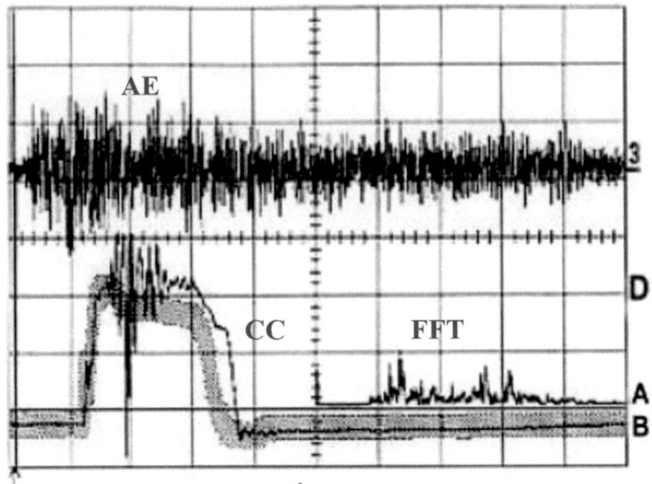

Figure 9.
Magnetron under abnormal operation with pulse voltage of 19.5 kV.

(zoom trace) and is shown superimposed on the current pulse. Channel A shows the magnitude FFT of the acoustic emission signal from channel 3. The amplitude of the AE signal decreased considerably during the detection of an anomalous pulse when compared with the AE under normal operation.

The experiment was repeated again using a type S9208A AE transducer. All parameters were initially set to normal values (filament primary at 115 V, pulse voltage at 15 kV, pulse rate at 1000 pps, and all RF power delivered to the matched load). Abnormal conditions were then obtained by gradually reducing the oscillator filament primary voltage from 115 V to zero. **Figures 10** and **11** show the results of this experiment under normal and abnormal conditions, respectively.

Once again, the amplitude of the acoustic emission signal decreased considerably during the detection of an anomalous pulse. Similar behavior was observed when the magnetron was stressed under different conditions. The system demonstrated the ability to detect anomalous pulses under different stress conditions. Further details of the experiments can be found in Ref. [7].

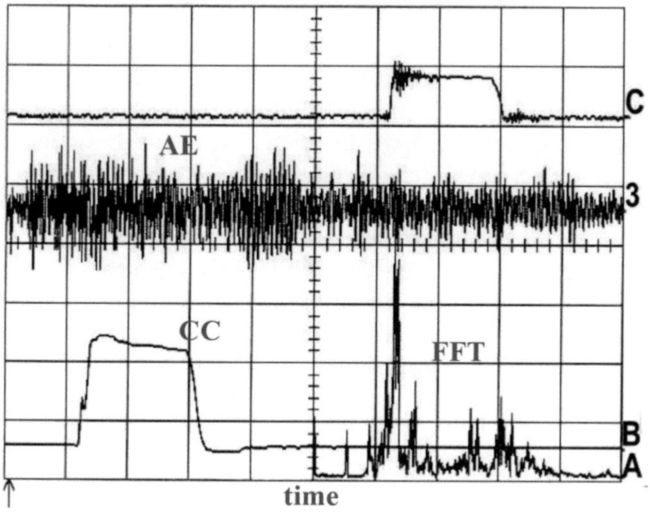

Figure 10.
Magnetron under normal operation with filament voltage of 115 V.

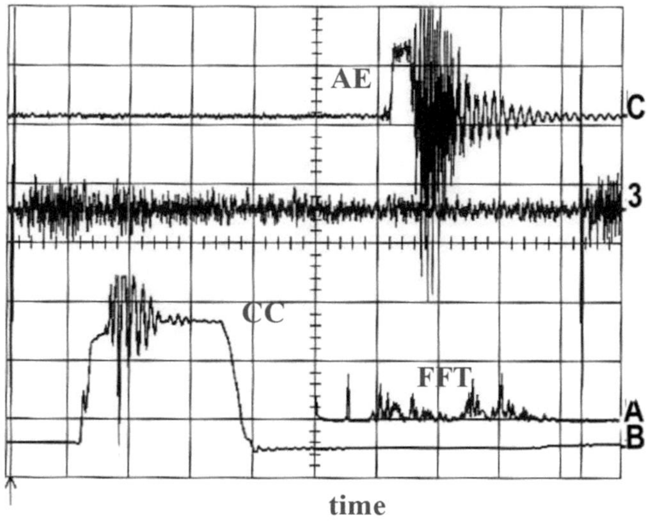

Figure 11.
Magnetron under abnormal operation with filament voltage of 0 V.

6. Integration with ICAS

The next step was to integrate the laboratory nondestructive AE test system into field applications for performance monitoring of a high-power microwave radar tube [8]. Currently, the Navy is using ICAS version 4.11 (IDAX Inc. Norfolk, VA) for the performance monitoring of mechanical systems on ships. The Integrated Condition Assessment System (ICAS) is a 32-bit Microsoft Windows NT based plant data analysis and integration tool. It is a predictive maintenance program that combines state-of-the-art performance monitoring techniques with computerized maintenance management. ICAS provides data acquisition and display, equipment analysis, diagnostic recommendations, and decision support information to plant operators and maintenance personnel. The system also provides user-defined performance alarms that alert the operator to machine problems. It provides the hybrid diagnostic system (HDS) diagnostic advisories that assist in diagnosing approaching failures and initiating the restoring process. The hybrid intelligent system is a fault-modeling environment that comprises both crisp logic and fuzzy logic rules.

The integration with ICAS required substantial electronics development in order to process, into a form compatible with ICAS, the acoustic emission and current signals collected during the normal and abnormal functioning of high-power radar tubes. An OPTO22 SNAP B3000 BRAIN unit was utilized to collect the data from the current and acoustic sensors. Two interface circuits were developed, one for each type of sensor, to interface with the ICAS software. **Figures 12** and **13** show the block diagrams for the developed electronics. The details of the electronics are beyond the scope of this paper. Further details on the ICAS interface and the electronics designed for this effort can be found in Ref. [8].

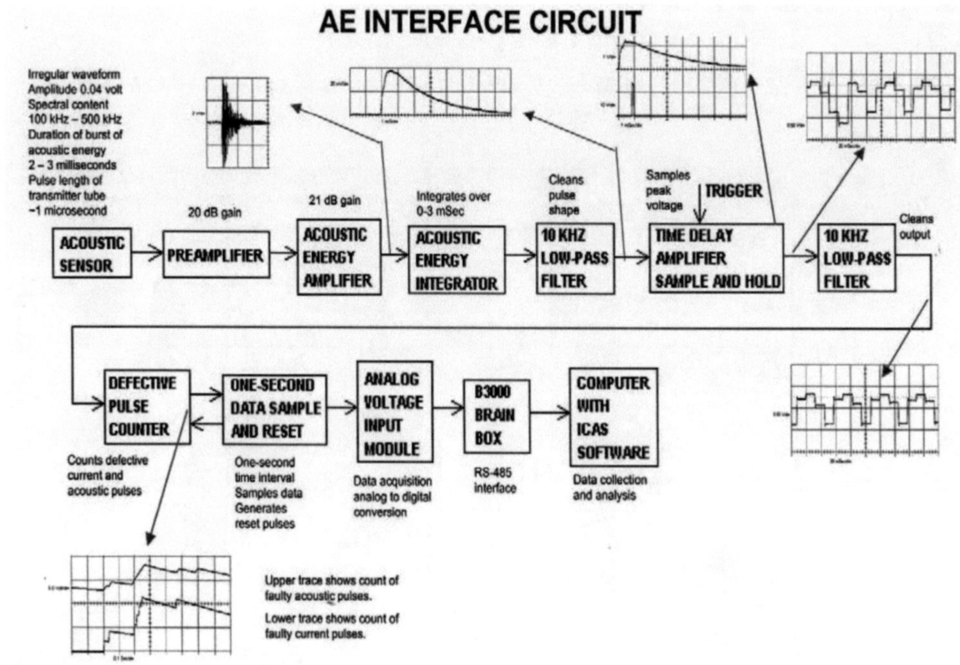

Figure 12.
Interface circuit for acoustic emission sensor.

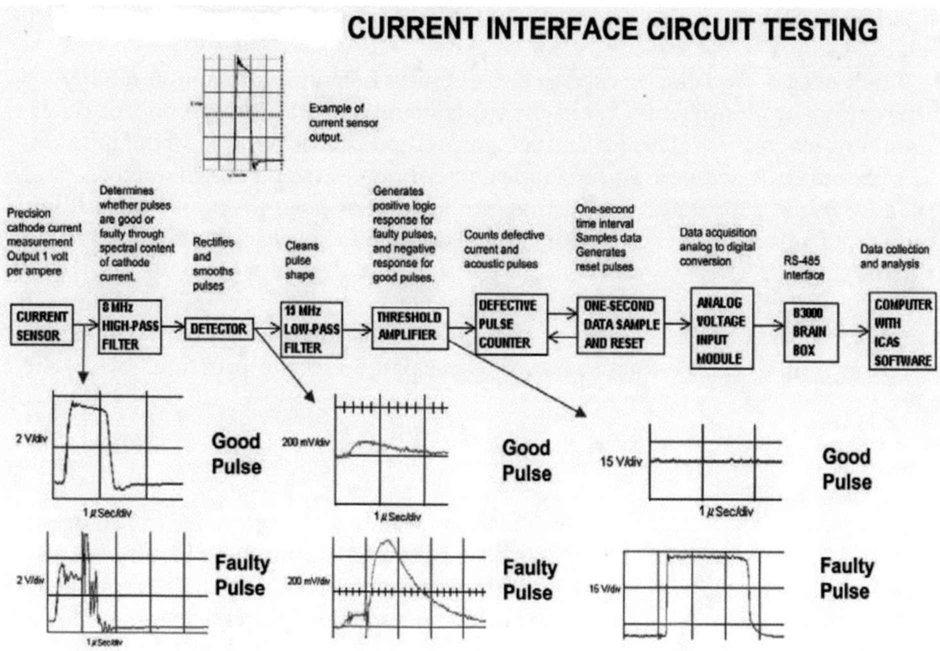

Figure 13.
Interface circuit for cathode current sensor.

7. Virtual sensor

One of the main advantages of the ICAS software is that it allows mathematical manipulation and combination of inputs from real sensors. This is essential in the development of what has been defined in this report as a virtual sensor. The generalized concept behind the virtual sensor is that a characteristic signature of a failure, denoted by F, is a function of more than one parameter $(x_1, x_2, x_3,\ldots,x_i)$. Thus, the characteristic signature of the failure, F, can be written mathematically as

$$F = F(x_1, x_2, x_3, \ldots, x_i) = \sum_{i=1}^{n}\sum_{j=1}^{n}\sum_{k=0}^{\infty}\sum_{l=0}^{\infty} C_{ijkl} X_i^k X_j^l \tag{3}$$

where C_{ijkl} are calculation coefficients and X^k and X^l correspond to each of n number of sensor parameters. The function F can generally be represented by a polynomial expansion.

As an example, if the failure is a function of just two parameters, then

$$F = F(x_1, x_2) = a + bx_1 + cx_2 + dx_1x_2 + ex_1^2x_2 + fx_1x_2 + \ldots \tag{4}$$

where the coefficients C_{ijkl} are represented by a, b, c, d, ..., and the functional parameters X^k are represented by x_1 and x_2. Note that this expansion does not specifically require that the parameters have the same units or appear correlated at first appearance. Monitoring the value of F will therefore provide a measure of the system status and/or identify or differentiate failures. Using the virtual sensor formalism allows one to concatenate sensor information to provide more information than normally derived from either sensor alone or used in normal combination. An alternative approach to simply combining sensor data in the manner just described would be to define specific ranges of values for each sensor parameter and assign weighted values to each specified range.

Single sensor data can be used to detect faulty behavior but cannot readily differentiate or identify failures in the trigger sources, microwave tube, or other modulator electronics. This virtual sensor method can be applied to the monitoring of anomalous acoustic and cathode current pulses which are characteristic of a failed RF pulse. More specifically, this method can be used to count the number of anomalous pulses from each one of the sensor interfaces described in Section 4 and **Figures 12** and **13**.

To illustrate the advantages of a virtual sensor, consider the combined failure function, F, for a particular placement of an acoustic emission sensor whose parameter is represented by E_{AE} and a current sensor whose parameter is represented by I_c, where

$$F = |I_C| + \int_0^t E_{AE}(t)dt. \quad (5)$$

A virtual sensor for this function which represents the magnitude of the current pulse and the integrated AE energy can be used to add the number of faulty counts from the two real sensors producing a virtual sensor output. Long-term trends and analysis can be used to characterize the behavior and identify trend signatures for different types of microwave tubes. This synergistic effect of virtual sensing adds diagnostic and, more importantly, prognostic capabilities to the ICAS or any other monitoring system. The failure function, F, can be adapted to the needs and complexity of any system and can be defined to extract specific information required from that system.

This technique was demonstrated on a magnetron tube (2J55). The experiment was conducted at the Microwave Tubes Built-In Test Project laboratory at SPAWAR Systems Center, San Diego. The upper section of **Figure 14** shows an ICAS screen capture of the cathode current (green) and acoustic emission (yellow) faulty pulse counts. The lower section shows the virtual sensor outputs where two failure functions, F1 and F2, have been defined and measured. Function F1 = x−y (yellow), where the difference between the outputs of the two real sensors represents virtual sensor 1, with x = cathode current and y = acoustic emission; and function F2 = x*y (green), where the product of the outputs of the

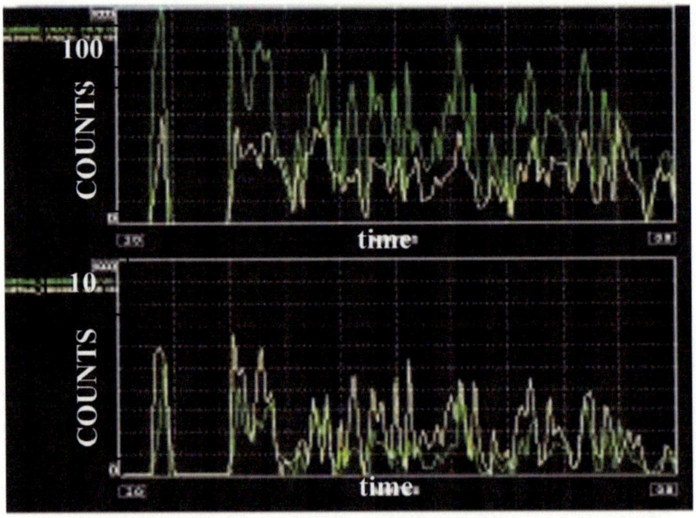

Figure 14.
Cathode current and acoustic emission faulty pulses and failure functions.

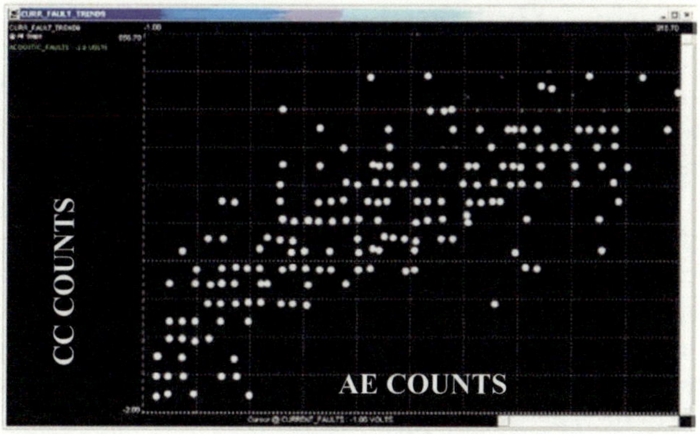

Figure 15.
Graph of faulty cathode current counts versus faulty acoustic emission counts.

two real sensors represents virtual sensor 2. These figures represent a 2-minute span of data collected during a period where the microwave tube was being stressed to produce faulty RF pulses.

The ICAS trending features were also demonstrated in this experiment. **Figure 15** shows a trend of the current faults count (vertical axis) versus the acoustic faults count (horizontal axis). This trend forms a band with few scattered points, indicating a strong correlation between both types of faulty counts and confirming the results obtained in previous experiments.

8. Conclusion

The experimental results presented in this paper demonstrate the use of advanced acoustic emission techniques as a nondestructive testing method for in-situ performance monitoring of high-power radar tubes such as pulsed magnetrons, TWTs, and klystrons. It was shown experimentally that changes in the amplitude and frequency content of the cathode current pulses are strongly correlated to changes in acoustic emission pulse energy both under normal and stressed operational conditions. The necessary electronics were developed to successfully interface the outputs of the current and acoustic emission sensors with the integrated condition assessment system (ICAS) used by the U.S. Navy. ICAS was used to demonstrate the use of virtual sensors where data from real sensors are combined into a failure functions F and captured in trend and single sensor format. In summary, this technique has demonstrated the unique ability to monitor, detect, and identify microwave tube performance. It has also demonstrated the ability to be utilized as a diagnostic tool by looking at long-term performance trends. More experimental research is needed to identify particular trends and signatures for the behavior of different types of microwave tubes under different circumstances to provide a fully prognostic capability for microwave tube systems.

Acknowledgements

This work was supported by the Office of Naval Research (ONR) under the auspices of Dr. Phillip Abraham (ONR-331) and Dr. Ignacio Perez (ONR-332); and in part by ONR's American Society for Engineering Education (ASEE) Summer Faculty Program, managed by Mr. Timothy Manicom.

Biographies and contact information quantum physics

Ayax D. Ramirez joined SPAWAR Systems Center, Diego in 1997. He received his M.S. and B.S. in Physics from San Diego State University in 1991 and 1984, respectively. Since 1984, he has performed research and development in the areas of laser physics, photonics, micro-sensors, excimer laser materials processing, and, more recently, in nondestructive detection techniques in high-power microwave tubes. Prior to working at SSC-SD, he worked in industry as a researcher for Science Applications International Corporation (SAIC). He holds numerous patents and technical publications in his field. He is a member of the American Physical Society and an adjunct professor at Southwestern College and Grossmont College, and he can be reached at ayax.ramirez@navy.mil.

Stephen D. Russell, Head of the Advanced Technology Branch at SPAWAR Systems Center, received his Ph.D. and M.S. in Physics from The University of Michigan in 1986 and 1981, respectively. He received his B.S. in Physics and Earth & Space Science from the State University of New York at Stony Brook in 1979. He has worked at SPAWAR Systems Center (and its predecessor organizations) since 1986 performing research and development in the fields of micro-sensors, photonic devices and systems, microelectronics, excimer laser processing, novel materials, and display technologies. Prior to joining SPAWAR Systems Center, he worked as a research assistant at The University of Michigan (1979–1986) and at Brookhaven National Laboratory (1977–1978). He has published numerous technical articles and has more than 60 U.S. and foreign patents issued or pending. He is a member of the American Physical Society and Optical Society of America and also serves as a consultant and technology agent for the Missile Defense Agency (MDA) and Defense Advanced Research Projects Agency (DARPA). He is a 2001 recipient of the Federal Laboratory Consortium Award for Excellence in Technology Transfer and holds two NASA Space Act Awards. He can be reached at stephen.russell@navy.mil.

David W. Brock, a Scientist in the Antennas and RF Technology Branch at SPAWAR Systems Center, received his Ph.D. in Physics from the University of Texas at Austin in 1981 and his M.S. in Physics from the University of Colorado at Boulder in 1975. He received B.S. degree in Physics and Mathematics at the New Mexico Institute of Mining and Technology in 1972. Prior to joining SPAWAR Systems Center, Dr. Brock worked as a research assistant at the University of Texas (1977–1981) and the University of Colorado (1975). He has worked at SPAWAR Systems Center since 1981 initially in the High Power Microwave Tube Branch and later in the Microwave/Millimeter Wave Technology Branch. He has carried out research work in areas including high power microwave tubes, the design and building of radar systems, microwave receivers, and antennas. He has published several technical articles, and has seventeen patents issued or pending. Dr. Brock can be reached at dbrock@spawar.navy.mil.

Narayan R. Joshi, faculty at the Welding Technology Department of North Harris College, received his M.Sc. in Electronics from Poona University, India. He received his M.S. and Ph.D. in Mechanics and Materials Science from Johns Hopkins University. He has worked as the Head of the dual Engineering Degree at the University of Maryland, Baltimore; Lamar University; and Prairie View A&M University. His academic teachings include elastic wave propagation, Materials Science, Metallurgy, Mechanical Vibrations, and Advanced Non-Destructive Evaluation Methods. His experimental research includes Ultrasonics, Acoustic Emission, X-ray diffraction and Optical Materials (Lasers without Inversion). Mr. Joshi has worked for General Electric, Oak Ridge National Laboratories, Southwest Research Institute, and Texas Instruments. He has about 55 research publications to his credit. He is a fellow of American Society for Non-Destructive Testing and a member of the American Society of Mechanical Engineers. He holds several patents issued and pending in his field, and he can be reached at giravani@juno.com.

Author details

Ayax D. Ramirez[1*], Stephen D. Russell[1], David W. Brock[1] and Narayan R. Joshi[2]

1 Space and Naval Warfare Systems Center, Pacific, San Diego, CA, USA

2 North Harris College, Houston, TX, USA

*Address all correspondence to: ayax.ramirez@navy.mil

References

[1] Leybovich A, Mark Ferdinand R. Sputtering target arc detection using acoustic emission. Materials Evaluation. Nov 2001;**59**:1331-1335

[2] Sakoda T, Nieda H, Ando, K. Characteristic of elastic waves caused by corona discharges in an oil-immersed pole transformer. IEEE Transactions on Dielectrics and Electrical Insulation. Apr 2001;**8**(2):276-283

[3] Boczar T. Results of Time-Frequency Analysis of Acoustic Emission Pulses Generated by Surface Partial Discharges in Air, Physics and Chemistry of Solid State. 2004;**5**(4):849-858

[4] Boczar T. Identification of fundamental forms of partial discharges based on the results of frequency analysis of their acoustic emission. Journal of Acoustic Emission. Jul-Dec 1999;**17**(3-4):S7-S12

[5] Theobald P et al. Reference source for the calibration of acoustic emission measurement. In: IEEE Instrumentation and Measurement Technology Conference; 21-23 May 2001; Budapest, Hungary. 2001. pp. 412-416

[6] Joshi NR, Brock DW, Russell SD, Lasher ME, Kasa SD. Built-in Test for High-Power Microwave Tubes Using Acoustic Emission Technology, SPAWAR Systems Center, San Diego (SSC-SD) Technical Note 1816. Sep 2001

[7] Joshi NR, Brock DW, Russell SD, Kasa SD, Garcia GA. Acoustic Emission Technology for Radar Tubes, SPAWAR Systems Center, San Diego (SSC-SD) Technical Note 1810. Aug 2000

[8] Joshi NR, Ramirez AD, Brock DW, Russell SD. Monitoring of High-Power Microwave Tube Systems Using the Integrated Condition Assessment System (ICAS), SPAWAR Systems Center, San Diego (SSC-SD) Technical Report 1885. Jul 2002

Section 2

Sound Propagation in Complex/Porous Materials

Chapter 4

Tortuosity Perturbations Induced by Defects in Porous Media

Fatma Graja and Claude Depollier

Abstract

In this chapter, we describe the effects of defects in a homogeneous saturated porous medium. Defects are modelized by inclusions which disturb the motion of the viscous fluid flowing in the pore space of the medium. The seepage rate of the fluid in the host medium and in the inclusion is given by the Darcy's law. Disturbances thus produced modify the shape of the stream lines from which we establish the tortuosity induced by the defects and its implications on the acoustic waves propagation in saturated porous media.

Keywords: tortuosity, defects, porous media, refractive index

1. Introduction

Among the essential physical parameters to describe the microstructure of porous media, tortuosity is one of the most important parameters. For a review, we can refer to the paper of Ghanbarian et al. [1].

Tortuosity was introduced as a correction to the permeability of Kozeny's model [2] of porous media defined by the Darcy's law relating the fluidic characteristics and pore space of the medium [3]:

$$\mathbf{v} = -\frac{k}{\eta}\nabla p, \tag{1}$$

where $\mathbf{v}$ is the seepage rate of the fluid, η the viscosity coefficient of the fluid, ∇p is the pressure gradient applied to the medium, and k is its permeability. The Kozeny's model was developed in the framework of straight and parallel streamlines in porous media. Carman has generalized it to neither straight nor parallel streamlines by introducing the hydraulic tortuosity τ defined by:

$$\tau = \frac{<\lambda>}{L}. \tag{2}$$

When a fluid flows through a porous medium from point A to point B distant from L (Euclidean distance) (**Figure 1**), it follows different paths whose mean length is $<\lambda>$, where λ is the length of the different paths connecting these two points. In isotropic media, the tortuosity is a scalar number greater than unit ($<\lambda> \geq L$), whereas for the low porous media, its values may be greater than 2; they range from 1 to 2 for high porosity media such as fibrous materials and some plastic foams.

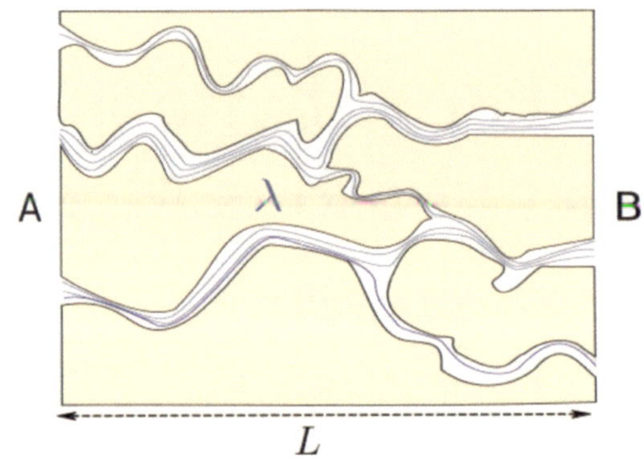

Figure 1.
Some tortuous paths through a porous medium.

The lengthening of the field paths in porous media due to tortuosity does not only occur in the flow of fluids in porous media, but is a more general result. So we meet this concept in processes such as transport phenomena, particles diffusion, electric conductivity, or wave propagation in fluid saturated porous media. Researchers have thus developed many theoretical models adapted to their concerns to introduce the tortuosity, leading to unrelated definitions of this concept. For instance, Saomoto and Katgiri [4] presented numerical simulations to compare hydraulic and electrical tortuosities. Thus, using numerical models of fluid flow and electric conduction in same media, i.e., with the same local solid phase arrangements, the authors show that while electrical tortuosity remains close to the unit whatever the porosity and the shape of the grains, the stream lines of hydraulic flow are much more concentrated in some parts of the medium, leading to a much greater tortuosity.

This example shows that although the physical meaning of this parameter is obvious, in practice, it is not consistent and its treatment is often misleading. The conclusion that emerges from these observations is that tortuosity should not be viewed as an intrinsic parameter of the environment in which the transport process develops, but rather as a property of this process. This partly explains why there are different definitions of tortuosity, each with its own interpretation.

In acoustics of porous media, tortuosity has been introduced to take into account the frequency dependence of viscous and thermal interactions of fluid motion with the walls of pores. In [5], Johnson uses it to renormalize the fluid density ρ_f. When the viscous skin depth is much larger than the characteristic dimensions of the pore, Lafarge et al. [6] have shown that the density of the fluid is equal to $\rho_f \tau_0$, where τ_0 is the static tortuosity for a constant flow ($\omega = 0$) defined by:

$$\tau_0 = \frac{<\mathbf{v}^2>}{<\mathbf{v}>^2}, \tag{3}$$

where $<.>$ denotes averaging over the pore fluid volume V_f. Thereafter, in this chapter, we adopt this definition of tortuosity.

Through the definition (3), we see that the tortuosity is given as soon as the permeability of the porous medium is known in each of its points. As it is well known, many factors can affect the fluid flow in porous media, including pore shape, distribution of their radii, and Reynolds number to name a few. It follows that the presence of defects in an initially homogeneous medium (for instance, a local change of an intrinsic parameter) can be an important disturbance of the fluid motion, the result being a modification of the shape of the streamlines.

Taking into account the presence of defects that change the permeability of the porous medium leads to the notion of effective permeability (k_{eff}). In general, the k_{eff} value is not unique but depends on the chosen model for the homogenization of the porous medium. The homogenization process only makes sense for lower scales than the spatial variations of incident excitation, which therefore justifies that mobility is calculated for a low-frequency filtration rate (quasi-static regime). These considerations lead us to be interested only in the instantaneous individual response of defects to external solicitations. Since in our case only media with low levels of defect are considered, it is legitimate to ignore their mutual interactions.

The present chapter is organized as follows. Section 2 describes the mathematical model of the defects and gives the solution of the fluid flow in the presence of homogeneous and layered spherical and ellipsoidal defects. Then, the results are generalized to anisotropic defects. Finally, the hydraulic polarizability is introduced. Section 3 is relative to tortuosity. The expression of effective mobility is given for some particular defects. The induced tortuosity is deduced from the previous results and its effects on the wave propagation are given.

2. Defect model

In this chapter, what is called defect is a local change of permeability k. Such a change is due, for instance, to variations in porosity in the microstructure of the medium. In this chapter, a defect is modelized as a porous inclusion Ω characterized by its shape and own parameters: intrinsic permeability $k^{(i)}$ and porosity $\phi^{(i)}$. Intrinsic permeability is expressed in darcy: $1D = 0.97 \times 10^{-12}\mathrm{m}^2$. The porous media we are interested in have permeabilities of the order of $10D$. Moreover, it is supposed that the fluid saturating the inclusion Ω is the same (with viscosity coefficient η) as that flowing in the porous medium. Thereafter the mobility of the fluid defined by $\kappa = k/\eta$ is used. This notion combines one property of the porous medium (permeability) with one property of the fluid (viscosity). The inclusion is embedded in a porous medium with porosity $\phi^{(o)}$ and permeability $k^{(o)}$. The saturating fluid is subject to action of a uniform pressure gradient $\nabla p^{(0)}$. In the sequel, we use indifferently the words defect or inclusion.

2.1 Mathematical formulation

When the fluid flows through the porous medium, its motion is perturbed by the defects in the microstructure of the medium. Within the porous medium, the velocity $\mathbf{v}$ and the pressure gradient ∇p are related by the Darcy's law:

$$\mathbf{v}^{(m)}(x) = \frac{k^{(m)}(x)}{\eta} \nabla p^{(m)}(x), \tag{4}$$

where $m = i$ if $x \in \Omega$ and $m = o$ if x is in the host medium ($x \notin \Omega$). These equations are subject to the following boundary conditions on $\partial\Omega$:

- continuity of fluid flow

$$\phi^{(o)}\mathbf{v}^{(o)}.\mathbf{n}^{(o)} = \phi^{(i)}\mathbf{v}^{(i)}.\mathbf{n}^{(i)}, \tag{5}$$

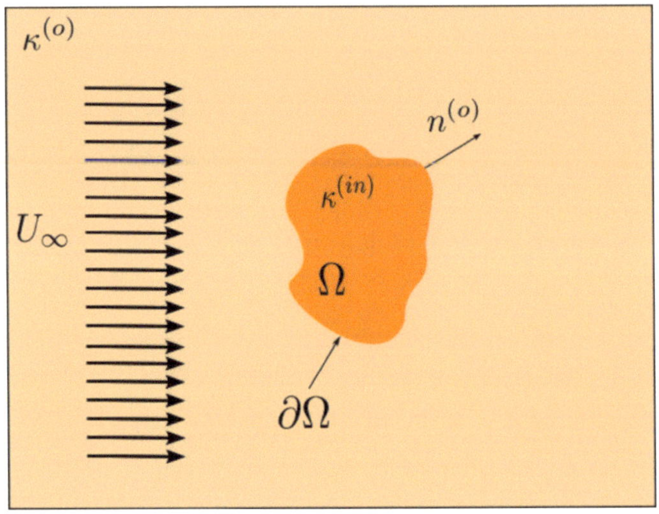

Figure 2.
Oriented inclusion in a porous medium.

- continuity of normal stress component

$$\tau_{ij}^{(o)} n_j^{(o)} = \tau_{ij}^{(i)} n_j^{(i)} \tag{6}$$

where $\mathbf{n}^{(o)}$ and $\mathbf{n}^{(i)}$ are unit vectors perpendicular to the interface. The hypothesis that Darcy's law governs the dynamics of the flow of fluid in a porous excludes inclusions filled with fluid. Indeed, within such inclusions, the movement of the fluid is governed by the Navier-Stokes equations, which are impossible to reconcile with the law of Darcy in the porous medium with the available boundary conditions [7]. **Figure 2** represents an oriented inclusion in a fluid in motion.

A porous medium of infinite extension is considered in which a viscous fluid flows at a constant uniform velocity U_∞ under the action of the pressure gradient along the Ox axis. We want to determine the local changes of the fluid velocity when defects are present in the medium.

In the following, we give the solutions of Eqs. (4)–(6) for some particular defects in such situation. Analytical solutions are possible for homogeneous spherical and ellipsoidal inclusions. We show that the most important characteristic of these inclusions is their hydraulic dipole moment. For layered defects, i.e., when their permeability is a piecewise constant function, we give a matrix-based method to get their dipolar moment.

2.2 Isotropic homogeneous defect

In this section, we assume that the background and the defect (embedded inclusion) are homogeneous each with its own parameters: porosity $\phi^{(o)}$ and $\phi^{(i)}$ and permeability $k^{(o)}$ and $k^{(i)}$ which have constant values. A static incident pressure with a constant gradient along Ox axis is applied to this system. What we seek is the pressure perturbation produced by the defect acting as a scatter and the expressions of the resulting seepage rate of the fluid inside and outside the porous inclusion.

2.2.1 Spherical defect

The simplest type of inclusion is the homogeneous spherical one, and we consider a porous sphere of radius $r = a$, centered at the origin of axes with a constant permeability $k^{(o)}$. Using the spherical coordinates (r, θ, φ), Eqs. (4)–(6) become

$$\mathbf{v}^{(m)} = \frac{k^{(m)}}{\eta} \nabla p^{(m)} \quad \text{where} \begin{cases} m = i & \text{if } r < a \\ m = o & \text{if } r > a \end{cases} \tag{7}$$

with the boundary conditions at $r = a$:

$$\phi^{(i)} v_r{}^{(i)}(r = a) = \phi^{(i)} v_r{}^{(i)}(r = a), \qquad \tau_{rr}^{(o)}\big|_{r=a} = \tau_{rr}^{(i)}\big|_{r=a} \quad \forall\theta, \tag{8}$$

where

$$\tau_{rr} = \phi\left(-p + 2\eta \frac{\partial v_r}{\partial r}\right), \tag{9}$$

v_r being the radial component of the velocity. For an incompressible fluid, Eq. (4) becomes

$$\Delta p^{(m)} = 0 \quad m = i, o. \tag{10}$$

In spherical coordinates, this equation is written as:

$$\Delta p = \frac{\partial}{\partial r}\left(r^2 \frac{\partial p}{\partial r}\right) + \frac{1}{\sin\theta} \frac{\partial p}{\partial \theta}\left(\sin\theta \frac{\partial p}{\partial \theta}\right) + \frac{1}{\sin^2\theta} \frac{\partial^2 p}{\partial^2 \varphi}. \tag{11}$$

The spherical symmetry of the problem (**Figure 3**) implies that the solution does not depend on φ. It follows that its solution is:

$$p^{(m)}(r, \theta) = \sum_l \left(A_l^{(m)} r^l + B_l^{(m)} r^{-(l+1)}\right) P_l(\cos\theta), \tag{12}$$

$P_l(x)$ being the Legendre polynomial of degree l. The coefficients $A_l^{(m)}$ and $B_l^{(m)}$ are related by the boundary conditions (8), those at $r = 0$ and when $r \to \infty$. Inside the inclusion, the pressure must be finite at $r = 0$. This condition leads to $B_l^{(i)} = 0$ for all l. So $p^{(i)}$ becomes:

$$p^{(i)}(r, \theta) = \sum_l A_l^{(i)} r^l P_l(\cos\theta). \tag{13}$$

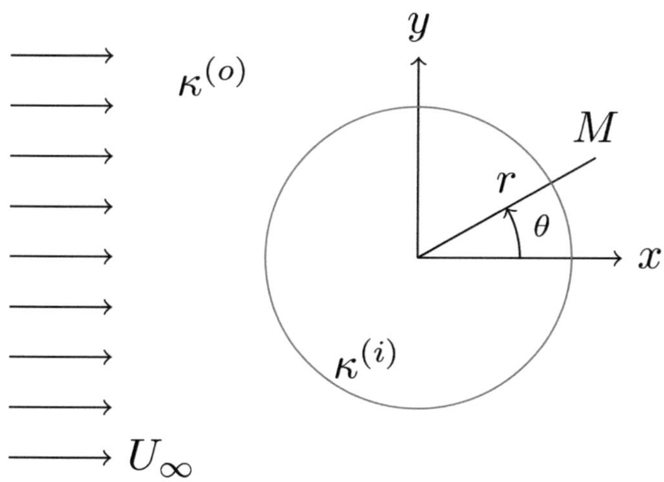

Figure 3.
Spherical inclusion in a fluid flow.

Outside, far from the inclusion, the pressure is:

$$p^{(o)} \sim -\frac{1}{\kappa^{(o)}} U_{\infty} r cos\theta. \tag{14}$$

It follows that the condition

$$p^{(o)}(r=a,\theta) = p^{(i)}(r=a,\theta) \quad \forall\theta, \tag{15}$$

which implies that only the term with $l = 1$ remains in the sum. Thus, we get:

$$A_1^{(o)} = -\frac{1}{\kappa^{(o)}} U_{\infty}. \tag{16}$$

The expressions of the pressure are then:

$$p^{(i)} = A_1^{(i)} r cos\theta \quad \text{if } r < a, \tag{17}$$

$$p^{(o)} = -\frac{U_{\infty}}{\kappa^{(o)}} r cos\theta + B_1^{(o)} r^{-2} cos\theta \quad \text{if } r > a, \tag{18}$$

$A_1^{(i)}$ and $B_1^{(o)}$ been given by the conditions (8). Finally, these expressions are:

$$p^{(i)}(r,\theta) = -\frac{\phi^{(o)}}{\phi^{(i)}} \frac{U_{\infty}}{\kappa^{(o)}} \left(\frac{3 + 12\frac{k^{(o)}}{a^2}}{2 + \frac{k^{(i)}}{k^{(o)}} + 12\frac{k^{(i)}}{a^2}} \right) r\cos\theta, \quad r < a \tag{19}$$

$$p^{(o)}(r,\theta) = -\frac{U_{\infty}}{\kappa^{(o)}} r\cos\theta + \frac{U_{\infty}}{\kappa^{(o)}} \frac{a^3}{r^2} \left(\frac{\frac{k^{(i)}}{k^{(o)}} - 1}{2 + \frac{k^{(i)}}{k^{(o)}} + 12\frac{k^{(i)}}{a^2}} \right) \cos\theta, \quad r > a \tag{20}$$

For defects with radius $r \sim 10^{-2}, 10^{-3}$ m, the quantities $k^{(i)}/a^2$ and $k^{(o)}/a^2$ are very small compared to unit ($\simeq 10^{-5}, 10^{-6}$) and can be neglected.

The velocity is deduced from the pressure due to Darcy's law.

- The pressure inside the inclusion (19) describes a constant velocity field. When the porosities of host medium and inclusion are substantially equal, then the fluid velocity is uniform (constant and aligned with the applied pressure gradient) with value

$$v^{(i)} = U_{\infty} \frac{3}{1 + 2\frac{\kappa^{(o)}}{\kappa^{(i)}}}. \tag{21}$$

If $\kappa^{(i)} > \kappa^{(o)}$, then $v^{(i)} > U_{\infty}$. In this case, the fluid passes preferentially through the inclusion and in the vicinity of the inclusion, the streamlines in the host medium are curved toward the inclusion. We have the inverse conclusion if $\kappa^{(i)} < \kappa^{(o)}$. When $\kappa^{(o)} \to 0$, i.e., for low permeability $k^{(o)}$, then $v^{(i)} \sim 3U_{\infty}$. If $\kappa^{(i)} \to 0$, then $v^{(i)} \to 0$.

Eq. (21) can also be written in the form:

$$v^{(i)} = U_{\infty} \frac{\kappa^{(i)}}{\kappa^{(o)} + \frac{1}{3}(\kappa^{(i)} - \kappa^{(o)})} \tag{22}$$

which will be generalized for the ellipsoidal inclusion.

- Outside the inclusion, the pressure is the sum of the applied pressure plus a dipolar contribution due to a induced dipole centered at the origin, the dipolar moment $\mathcal{P}_{sph}$ of which is

$$\mathcal{P}_{sph} = 4\pi a^3 U_\infty \kappa^{(i)} \left(\frac{\frac{k^{(i)}}{k^{(o)}} - 1}{2 + \frac{k^{(i)}}{k^{(o)}}} \right). \tag{23}$$

The corresponding density of induced "hydraulic" surface charges is

$$\sigma_{sph}(\theta) = 4\pi a^3 U_\infty \kappa^{(i)} \left(\frac{\frac{k^{(i)}}{k^{(o)}} - 1}{2 + \frac{k^{(i)}}{k^{(o)}}} \right) \cos(\theta). \tag{24}$$

In the host medium $(r > a)$, the components of the seepage rate are:

$$v_r^{(o)} = -U_\infty \cos\theta \left(1 + 2\frac{a^3}{r^3} \left(\frac{\frac{\kappa^{(i)}}{\kappa^{(o)}} - 1}{2 + \frac{\kappa^{(i)}}{\kappa^{(o)}}} \right) \right), \tag{25}$$

$$v_\theta^{(o)} = U_\infty \sin\theta \left(1 - \frac{a^3}{r^3} \left(\frac{\frac{\kappa^{(i)}}{\kappa^{(o)}} - 1}{2 + \frac{\kappa^{(i)}}{\kappa^{(o)}}} \right) \right). \tag{26}$$

Figure 4 represents the seepage rate in the porous medium for $\kappa^{(i)} > \kappa^{(o)}$. **Figure 4a** shows the levels of the amplitude of the velocity, while **Figure 4b** shows its stream lines.

So the response developed by a defect when submitted to a pressure gradient is an induced dipole $\mathcal{P}$. When dealing with linear phenomena (low filtration speed), this response is proportional to its cause, the proportionality factor depending only on the shape of the defect, its volume, and on the ratio of its mobility to that of the host medium. In our case, the dipole moment is

$$\mathcal{P}_{sph} = \alpha(-U_\infty), \tag{27}$$

where α is a susceptibility which assesses the polarizability, i.e., the capacity of the porous inclusion to induce a dipole $\mathcal{P}_{sph}$ under the action of excitation $-U_\infty/\kappa^{(o)}$. For a spherical inclusion of volume V, the susceptibility α is:

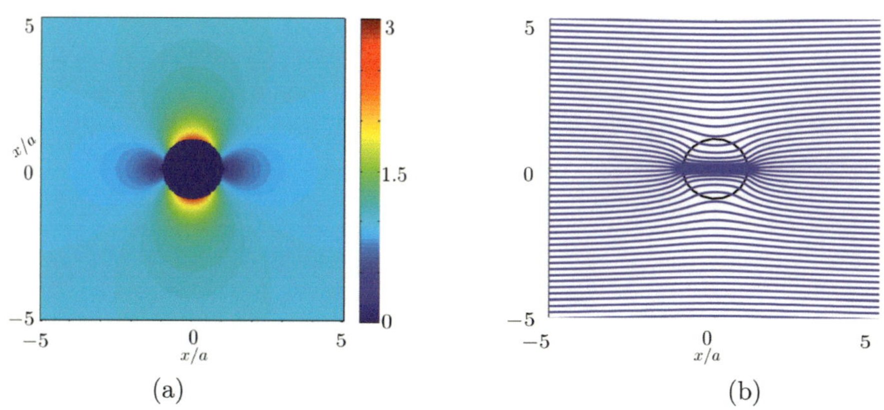

Figure 4.
Fluid flow through an inclusion in a porous medium: (a) velocity levels and (b) streamlines ($\kappa^{(i)}/\kappa^{(o)} > 1$).

$$\alpha = 3V\kappa^{(i)}\left(\frac{\frac{k^{(i)}}{k^{(o)}} - 1}{2 + \frac{k^{(i)}}{k^{(o)}}}\right). \tag{28}$$

Often, the quantity $\chi = \alpha/\kappa^{(i)}V$ which does not depend on the volume of the inclusion is more relevant. Its variations as function of the ratio $\kappa = \kappa^{(i)}/\kappa^{(o)}$ are shown in **Figure** 5. They range from $-3/2$ when $\kappa^{(i)} << \kappa^{(o)}$ to 3 when $\kappa^{(i)} >> \kappa^{(o)}$. In the following, it is convenient to put the external pressure in the form:

$$p^{(o)}(r,\theta) = -\frac{U_\infty}{\kappa^{(o)}} r cos\theta + \frac{P_{sph}}{4\pi}\frac{cos\theta}{r^2} \tag{29}$$

2.2.2 Ellipsoidal defect

In addition to the interest that ellipsoidal inclusion has an exact analytical solution, its study (its study) allows us to understand the effects of the shape of the defects on the fluid motion. Indeed, ellipsoidal surface can be seen as a generic element of a set of volumes comprising the disc, the sphere, and the oblong shape (needle) and so the nonsphericity can be appreciated through the values of its polarizability. The general ellipsoidal inclusion having semiaxes a, b, and c aligned with the axes of the Cartesian coordinates system and centered at the origin is described by the following equation:

$$\frac{x^2}{u+a^2} + \frac{y^2}{u+b^2} + \frac{z^2}{u+c^2} = 1, \tag{30}$$

where x, y, and z are the position coordinates of any point on the surface of the ellipsoid. Eq. (30) has three roots ξ, η, and ζ which define the surfaces coordinates: surfaces with constant ξ are ellipsoids, while surfaces with constant η or ζ are hyperboloids. Surfaces of confocal ellipsoids are described by adjusting the scalar u. So, two ellipsoids defined by (30) with $u = u_1$ and $u = u_2$ are called confocal if their semi axes obey to the conditions

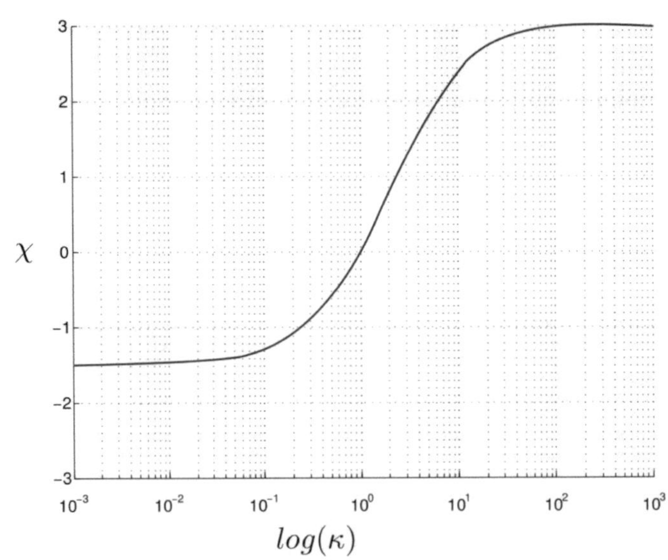

Figure 5.
Hydraulic polarizability $\chi = \alpha/\kappa^{(i)}V$ of a spherical inclusion vs log (k).

$$a_1^2 - a_2^2 = b_1^2 - b_2^2 = c_1^2 - c_2^2 \tag{31}$$

where a_i, b_i, and c_i are their respective semiaxes. Some results related to the ellipsoid are given in Appendix A.

Let $p(\mathbf{r})$ be the pressure with a constant gradient directed along the Ox axis applied to the porous medium (**Figure 6**). In absence of defect, its expression is:

$$p(\mathbf{r}) = p_0(x) = Ex = E\sqrt{\frac{(a^2+\xi)(a^2+\eta)(a^2+\zeta)}{(b^2-a^2)(c^2-a^2)}} \tag{32}$$

where $E = -U_\infty/\kappa^{(o)}$. In this relation, ξ, η, and ζ are the ellipsoidal coordinates given in Appendix A. Here, the field $p_0(x)$ can be viewed as an "incident pressure."

When the inclusion is embedded in the porous medium, it produces perturbations in the fluid motion giving rise to the "scattered" pressure. When the filtration rate is low, the scattering by the inclusion is a linear phenomenon, leading to the following expression of the scattered pressure:

$$p_{sc}(\mathbf{r}) = p_0(x)F(\xi). \tag{33}$$

where $F(\xi)$ is a proportional coefficient. For $p_{sc}(\mathbf{r})$ to be a solution of the Laplace equation, it must verify the differential equation:

$$\frac{d^2F}{d\xi^2} + \frac{dF}{d\xi}\frac{d}{d\xi}\left(\ln\left[R(\xi)\left(\xi+a^2\right)\right]\right) = 0 \tag{34}$$

where $R(\sigma) = (a^2+\sigma)\left(b^2+\sigma\right)(c^2+\sigma)$. $F(\xi)$ is then the sum of the two functions $F_1(\xi) + F_2(\xi)$, where $F_1(x) = A$ is a constant and $F_2(x)$ is

$$F_2(\xi) = \int_\xi^\infty \frac{d\sigma}{(\sigma+a^2)R(\sigma)}. \tag{35}$$

Thus, the pressure outside the inclusion is

$$p(\mathbf{r}) = p_0(x)\left[A - \frac{B}{2}\int_\xi^\infty \frac{d\sigma}{(\sigma+a^2)R(\sigma)}\right]. \tag{36}$$

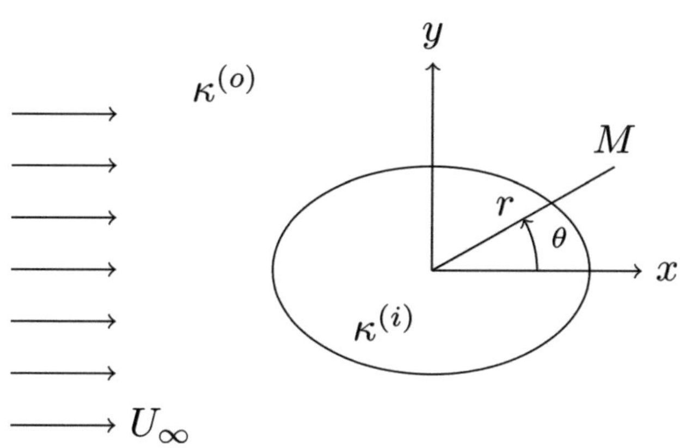

Figure 6.
Ellipsoidal inclusion in a fluid flow.

The two constants A and B are determined by the boundary conditions at the surface $\xi = 0$ of the inclusion:

1. continuity of fluid flow at $\xi = 0$:

$$\phi^{(e)}\kappa^{(e)}\frac{1}{h_\xi}\frac{\partial p^{(e)}}{\partial \xi} = \phi^{(i)}\kappa^{(i)}\frac{1}{h_\xi}\frac{\partial p^{(i)}}{\partial \xi}, \tag{37}$$

2. continuity of the stress component $\tau_{\xi\xi}$ in fluid at $\xi = 0$:

$$\phi^{(e)}\left(-p^{(e)} + 2\eta\frac{1}{h_\xi}\frac{\partial v_\xi^{(e)}}{\partial \xi}\right) = \phi^{(i)}\left(-p^{(i)} + 2\eta\frac{1}{h_\xi}\frac{\partial v_\xi^{(i)}}{\partial \xi}\right) \tag{38}$$

where v_ξ is the normal component of the seepage flow to the surface $\xi = 0$ given by the Darcy's law:

$$v_{\lambda_i} = -\kappa\frac{1}{h_{\lambda_i}}\partial_{\lambda_i}p, \quad \lambda_i = \xi, \eta, \zeta. \tag{39}$$

In these relations, the coefficients h_{λ_i} are the scale factors of ellipsoidal coordinates given in Appendix A. We obtain the values of the pressure such that:

1. within the inclusion:

$$p^{(i)}(x) = \frac{\kappa^{(o)}}{\kappa^{(o)} + N_x\left(\kappa^{(i)} - \kappa^{(o)}\right)}p_0(x), \tag{40}$$

2. outside the inclusion:

$$p^{(o)}(x) = p_0(x) + p_{sc}(x) = p_0(x) - \frac{abc\left(\kappa^{(i)} - \kappa^{(o)}\right)}{\kappa^{(o)} + N_x\left(\kappa^{(i)} - \kappa^{(o)}\right)}p_0(x)F_2(\xi). \tag{41}$$

Far from the center of the inclusion, $\xi \approx r^2$, the scattered pressure can be approximate by:

$$\begin{aligned} p_{sc}(x) &\approx \frac{abc\left(\kappa^{(i)} - \kappa^{(o)}\right)}{\kappa^{(o)} + N_x\left(\kappa^{(i)} - \kappa^{(o)}\right)}p_0(x)\int_{r^2}\frac{dr}{r^{5/2}} \\ &\approx \frac{abc\left(\kappa^{(i)} - \kappa^{(o)}\right)}{\kappa^{(o)} + N_x\left(\kappa^{(i)} - \kappa^{(o)}\right)}\frac{x}{3r^3}E. \end{aligned} \tag{42}$$

The right-hand side of Eq. (42) is the expression of pressure produced by a dipole aligned with the axis Ox. From the expression of the speed within the inclusion:

$$v^{(i)} = U_\infty\frac{\kappa^{(o)}}{\kappa^{(o)} + N_x\left(\kappa^{(i)} - \kappa^{(o)}\right)}, \tag{43}$$

the dipole moment is then

$$\mathcal{P}_{elli,a} = -VU_\infty\kappa^{(i)}\frac{\left(\kappa^{(i)} - \kappa^{(o)}\right)}{\kappa^{(o)} + N_x\left(\kappa^{(i)} - \kappa^{(o)}\right)}. \tag{44}$$

Here, V is the volume of the ellipsoidal inclusion. The factor N_x,

$$N_x = \frac{abc}{3}\int_0^{\infty} \frac{d\sigma}{(\sigma + a^2)\sqrt{\sigma + a^2})(\sigma + b^2)(\sigma + c^2)}, \tag{45}$$

describes how the dipole moment of the inclusion changes with its shape and its orientation in relation with the incident pressure field. The geometric parameters N_x, N_y, and N_z appear for the first time in hydrodynamics [8] to describe the disturbance brought by a solid immersed in an infinite fluid in uniform motion. Their values were computed by Stoner [9] and Osborn [10]. The name "depolarization factors" comes from electromagnetism (see, for example, Landau and Lifchitz [11]).

From Eq. (163), it is possible to find the values of the depolarization factors of some particular inclusions such as:

- spherical inclusion $N_x = N_y = N_z = 1/3$,
- inclusion in disc form (axis Oz) $N_x = N_y = 0, N_z = 1$, and
- oblong inclusion $N_x = N_y = 1/2, N_z = 0$.

Figure 7 is a plot of depolarization factors of a ellipsoidal inclusion having two equal semiaxes according to their ratio.

The expression of $p_{sc}(x)$ given by (42) is not exact since it is a result of the approximation $\xi \sim r^2$, i.e., far from the inclusion, where only the dipolar effects are relevant.

In the case of an ellipsoidal inclusion, the polarizability is no longer a scalar but is a tensor. Its eigenvalues are polarizabilities along the axes of the ellipsoid. So, we can write the dipole moment (44) as $\mathcal{P}_{elli,a} = \alpha_a U_\infty$, where α_a is the eigenvalue of the tensor polarizability along its principal direction Ox which defines the polarizability along this axis. In **Figure 8**, we depict the variations of the susceptibility χ as

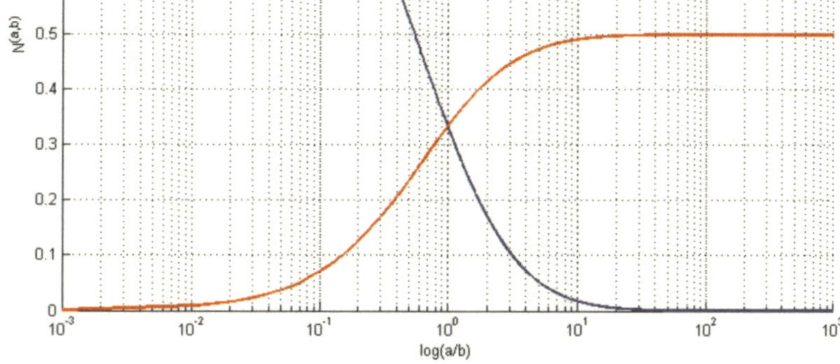

Figure 7.
Depolarization factors for a ellipsoidal inclusion with semiaxes (a, a, b) *as functions of* a/b.

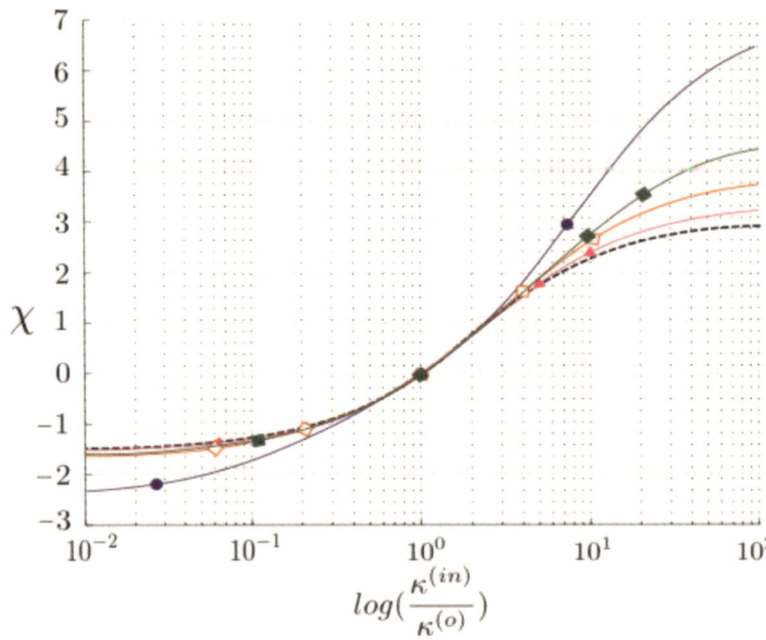

Figure 8.
Hydraulic polarizability $\chi = \alpha_x/V$ of an ellipsoidal inclusion vs. $\log\left(\kappa^{(i)}/\kappa^{(o)}\right)$ *for different values of the depolarization factors (N_x, N_y, N_z) (−•−: $N_x = 0.1, N_y = 0.1, N_z = 0.8$); (−⋄−: $N_x = 0.2, N_y = 0.2, N_z = 0.6$); (−◀−: $N_x = 0.4, N_y = 0.4, N_z = 0.2$); (−■−: $N_x = 0.45, N_y = 0.45, N_z = 0.1$); (− − −: $N_x = 1/3, N_y = 1/3, N_z = 1/3$).*

function of $\kappa^{(i)}/\kappa^{(o)}$ for different sets of the depolarization factors when incident pressure is along Ox.

The pressure outside the inclusion is then:

$$p^{(o)}(r,\theta) = -\frac{U_\infty}{\kappa^{(o)}} r\cos\theta + \frac{\mathcal{P}_{elli,\,a}}{4\pi}\frac{\cos\theta}{r^2}. \tag{46}$$

This relation is similar to (29), differing from it only by the expression of the dipole moment. The fundamental difference between the spherical and ellipsoidal inclusions is that the pressure scattered by the sphere contains only a dipolar field, whereas in strictness, the ellipsoid also scatters high-order multipolar fields. We can then deduce from this remark that the more the shape of the inclusion is distant from that of the sphere, the more the scattered pressure contains high-order multipolar terms.

The result obtained in (42) does not show these terms since the calculation of the integral F_2 is an approximate computation when $\xi/r \gg 1$. When we move away from the ellipsoidal inclusion, the multipolar terms of order greater than 2 decrease very quickly, leaving only the contribution of the incident pressure and the dipole term of the scattered pressure.

2.3 Inhomogeneous defect

In fact, defects rarely exhibit homogeneous structure. The parameter that characterizes the defect (in our case the permeability) is generally a variable varying according to a law which depends on the way in which the defect develops.

For the spherical defect, the simplest situation is the radial variation of the permeability. The fundamental difference between homogeneous and inhomogeneous spherical inclusions is that in the latter case, the velocity field loses its uniformity. The determination of the dipole moment requires a different approach from that previously developed. Two cases are considered: (i) the permeability is a piecewise constant function and (ii) the permeability is a continuously varying function.

2.3.1 Layered spherical defect

Consider an inhomogeneous sphere of porous medium embedded in a homogeneous host medium. We assume that the permeability of the sphere depends only on the radius and is a piecewise constant function, i.e., the sphere is a set of nested spherical layers. The permeability of the background medium is $k^{(o)}$, that of the outermost layer is $k^{(1)}$, and so on to the central sphere whose permeability is $k^{(i)}$.

To calculate the perturbation of the incident pressure due to the sphere seen as a scatter, we proceed the following: the pressure field is calculated in each layer of the sphere. The pressure field in the layer number n is related to those in the layer number $n+1$ and $n-1$ by the boundary conditions. It is assumed that the defect is a set of N concentric spherical layers in which the value of permeability is constant $k^{(n)}$. Let $\kappa^{(n)}$ denotes the ratio $k^{(n)}/\eta$ in the layer number n delimited by the spheres of radii r_n and r_{n+1} such that $r_n > r_{n+1}$. The core of the sphere has index $i = N$ (**Figure 9**). The determination of pressure and velocity in this type of inclusion consists in solving the Laplace equation $\Delta p^{(n)} = 0$ in each layer and in connecting the solutions using the boundary conditions: continuity of fluid flow and that of the radial component of the stress.

In the layer number n, the pressure is noted:

$$p^{(n)} = A^{(n)} r \cos\theta + B^{(n)} r^{-2} \cos\theta. \tag{47}$$

The coefficients A and B of two consecutive layers are connected by the following conditions in $r = r_{n+1}$:

$$\phi^{(n)} v_r^{(n)} (r = r_{n+1}) = \phi^{(n+1)} v_r^{(n+1)} (r = r_{n+1}), \tag{48}$$

$$\tau_{rr}^{(n)} (r = r_{n+1}) = \tau_{rr}^{(n+1)} (r = r_{n+1}). \tag{49}$$

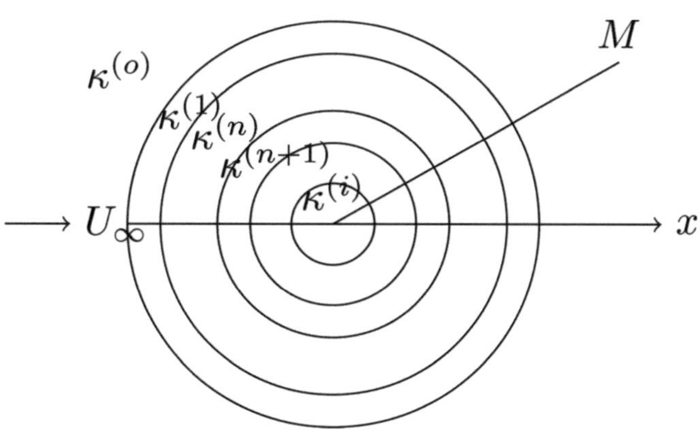

Figure 9.
Layered sphere.

Transfer matrix: The linearity of the problem makes it possible to write that the pairs of coefficients (A_n, B_n) and (A_{n+1}, B_{n+1}) are linked by a matrix equation such that:

$$\begin{pmatrix} A^{(n)} \\ B^{(n)} \end{pmatrix} = T_{n,n+1} \begin{pmatrix} A^{(n+1)} \\ B^{(n+1)} \end{pmatrix}, \tag{50}$$

where $T_{n,n+1}$ is the transfer matrix between the two consecutive layers n and $n+1$, the entries of which are:

$$T_{11} = \frac{\phi^{(n+1)}}{\phi^n} \left(\frac{k^{(n+1)}}{k^n} + 2 \frac{1 - \frac{k^{(n+1)}}{k^n}}{3 + 12k^n r_n^{-2}} \right), \tag{51}$$

$$T_{12} = 2\frac{\phi^{(n+1)}}{\phi^n} r_n^{-3} \left(-\frac{k^{n+1}}{k^n} + \frac{1 + 2\frac{k^{(n+1)}}{k^n} + 12k^{(n+1)} r_n^{-2}}{3 + 12k^n r_n^{-2}} \right), \tag{52}$$

$$T_{21} = \frac{\phi^{(n+1)}}{\phi^n} r_n^3 \left(\frac{1 - \frac{k^{(n+1)}}{k^n}}{3 + 12k^n r_n^{-2}} \right), \tag{53}$$

$$T_{22} = \frac{\phi^{(n+1)}}{\phi^n} \left(\frac{1 + 2\frac{k^{(n+1)}}{k^n} + 12k^{(n+1)} r_n^{-2}}{3 + 12k^n r_n^{-2}} \right). \tag{54}$$

From $A^0 = -U_\infty / \kappa^{(o)}$ and $B^{(i)} = 0$, it is possible to get the pressure in each layer of the sphere.

The effects of two consecutive layers are obtained by the product of the transfer matrices of each of these layers. So from the matrix equations:

$$\begin{pmatrix} A^{(n)} \\ B^{(n)} \end{pmatrix} = T_{n,n+1} \begin{pmatrix} A^{(n+1)} \\ B^{(n+1)} \end{pmatrix}, \tag{55}$$

$$\begin{pmatrix} A^{(n+1)} \\ B^{(n+1)} \end{pmatrix} = T_{n+1,n+2} \begin{pmatrix} A^{(n+2)} \\ B^{(n+2)} \end{pmatrix}, \tag{56}$$

we get:

$$\begin{pmatrix} A^{(n)} \\ B^{(n)} \end{pmatrix} = T_{n,n+2} \begin{pmatrix} A^{(n+2)} \\ B^{(n+2)} \end{pmatrix}, \tag{57}$$

where $T_{n,n+2} = T_{n,n+1} T_{n+1,n+2}$.

Scattering matrix: Another way of linking the coefficients A and B of two consecutive layers is the use of the scattering matrix $S_{n,n+1}$. The scattering matrix is sometimes more efficient than the transfer matrix for calculating the amplitudes of the waves reflected and transmitted by an object subjected to incident waves. It is such that:

$$\begin{pmatrix} B^{(n)} \\ A^{(n+1)} \end{pmatrix} = S_{n,n+1} \begin{pmatrix} A^{(n)} \\ B^{(n+1)} \end{pmatrix}. \tag{58}$$

Its entries are:

$$S_{n,n+1} = \frac{1}{\kappa^{(n+1)} + 2\kappa^{(n)}} \begin{pmatrix} \left(\kappa^{(n+1)} - \kappa^{(n)}\right) r_{n+1}^3 & 3\kappa^{(n+1)} \\ 3\kappa^{(n)} & 2\left(\kappa^{(n+1)} - \kappa^{(n)}\right) r_{n+1}^{-3} \end{pmatrix}. \tag{59}$$

The scattering matrix of two consecutive layers is given by their Redheffer product. From

$$\begin{pmatrix} B^{(n)} \\ A^{(n+1)} \end{pmatrix} = S_{n,n+1} \begin{pmatrix} A^{(n)} \\ B^{(n+1)} \end{pmatrix}, \tag{60}$$

$$\begin{pmatrix} B^{(n+1)} \\ A^{(n+2)} \end{pmatrix} = S_{n+1,n+2} \begin{pmatrix} A^{(n+1)} \\ B^{(n+2)} \end{pmatrix}, \tag{61}$$

we get:

$$\begin{pmatrix} B^{(n)} \\ A^{(n+2)} \end{pmatrix} = S_{n,n+2} \begin{pmatrix} A^{(n)} \\ B^{(n+2)} \end{pmatrix}, \tag{62}$$

where $S_{n,n+2} = S_{n,n+1}{}^{*} S_{n+1,n+2}$. In this relation, the right-hand side is the Redheffer star product. For more details about the Redheffer star product, we can refer to [12].

2.3.2 Spherical inclusion with continuously variable permeability

When κ is a continuous function of the variable r, the matrix Eq. (50) becomes the system of differential equations:

$$\frac{d}{dr}\left(\frac{r^3\kappa^2 dA/dr}{d\kappa/dr}\right) + r^2\kappa A = 0 \tag{63}$$

$$\frac{d}{dr}\left(\frac{\kappa^2 dB/dr}{r^3 d\kappa/dr}\right) - \frac{2\kappa}{r^4}B = 0. \tag{64}$$

2.3.3 Layered ellipsoidal inclusion

The generalization of the radial variation of the permeability of the spherical inclusion to the ellipsoidal requires that the permeability only depends on ξ. This is true in orthogonal directions at its surface. This condition entails that inside the ellipsoid, the strata are limited by confocal ellipsoidal surfaces $\xi = \xi_k$, i.e., having the same foci as the surface ξ; hence, their semiaxes are related by the following relations:

$$a^2 - a_k^2 = b^2 - b_k^2 = c^2 - c_k^2. \tag{65}$$

Consider a porous inhomogeneous ellipsoidal inclusion having the permeability $k^{(i)}$ embedded in a background medium of homogeneous mobility $\kappa^{(o)}$. We assume that the mobility of the inclusion is stratified, i.e., it is a constant piecewise function and each layer has its own mobility $\kappa^{(n)}$. The mobility in the outermost layer is $\kappa^{(1)}$, that of the next layer is $\kappa^{(2)}$, etc. to the central layer whose the mobilty is $\kappa^{(i)}$. The layers are limited by the confocal surfaces $\xi = \xi_k$ whose semiaxes obey (65) and are numbered from 1 to $i = N$ from the outside to the inside, such that the strata n and

$n+1$ have the common boundary $\xi = \xi_{n+1}$ (**Figure 10**). In each of these strata, the pressure is the solution of the Laplace equation given by:

$$p^{(j)}(\mathbf{r}) = -Ex\left[A^{(j)} - \frac{B^{(j)}}{2}\int_{\xi}^{\infty}\frac{d\sigma}{((\sigma+a^2)R^{(j)}(\sigma)}\right] \tag{66}$$

where

$$R^{(j)}(\sigma) = \left(a_j^2+\sigma\right)\left(b_j^2+\sigma\right)\left(c_j^2+\sigma\right). \tag{67}$$

In the layers n and $n+1$, the solutions of the Laplace equations are:

$$p^{(n)}(\mathbf{r}) = -Ex\left[A^{(n)} - \frac{B^{(n)}}{2}\int_{\xi}^{\infty}\frac{d\sigma}{((\sigma+a^2)R^{(n)}(\sigma)}\right], \tag{68}$$

$$p^{(n+1)}(\mathbf{r}) = -Ex\left[A^{(n+1)} - \frac{B^{(n+1)}}{2}\int_{\xi}^{\infty}\frac{d\sigma}{((\sigma+a^2)R^{(n+1)}(\sigma)}\right] \tag{69}$$

with the boundary condition at $\xi = \xi_{n+1}$:

$$\phi^{(n)}\frac{1}{h_{\xi(n)}}\frac{\partial p^{(n)}}{\partial\xi} = \phi^{(n+1)}\frac{1}{h_{\xi(n+1)}}\frac{\partial p^{(n+1)}}{\partial\xi}, \tag{70}$$

and

$$\phi^{(n)}\left(-p^{(n)} + 2\eta\frac{\partial v_{\xi}^{(n)}}{h_{\xi(n)}\partial\xi}\right) = \phi^{(n+1)}\left(-p^{(n+1)} + 2\eta\frac{\partial v_{\xi}^{(n+1)}}{h_{\xi(n+1)}\partial\xi}\right). \tag{71}$$

By proceeding in the same way as for the spherical cavity, one finds the transfer matrix and the scattering matrix.

Transfer matrix: The transfer matrix of the ellipsoidal inclusion is:

$$T_{n,n+1} = \frac{1}{\kappa^{(n)}}\begin{pmatrix} \kappa^{(n)} + N_{x,n+1}\left(\kappa^{(n+1)}-\kappa^{(n)}\right) & \dfrac{N_{x,n+1}(1-N_{x,n+1})}{a_{n+1}b_{n+1}b_{n+1}}\left(\kappa^{(n+1)}-\kappa^{(n)}\right) \\ a_{n+1}b_{n+1}c_{n+1}\left(\kappa^{(n+1)}-\kappa^{(n)}\right) & \kappa^{(n+1)} + N_{x,n+1}\left(\kappa^{(n)}-\kappa^{(n)}\right) \end{pmatrix}. \tag{72}$$

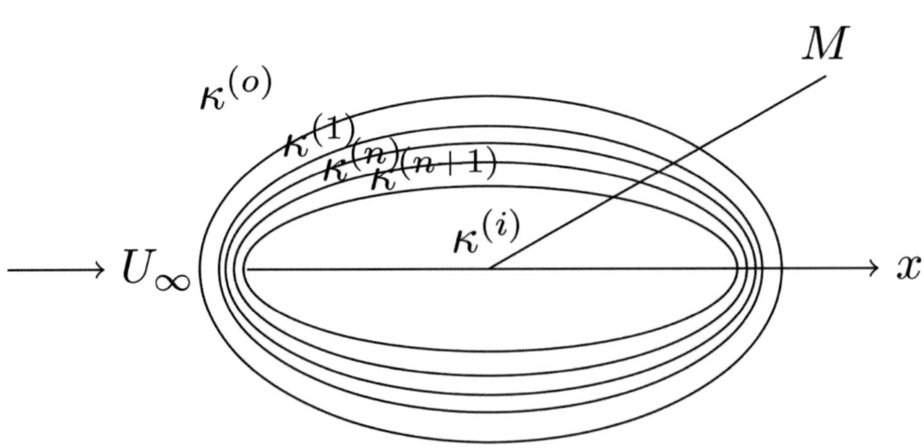

Figure 10.
Layered ellipsoidal inclusion.

Scattering matrix: The scattering matrix of the ellipsoidal inclusion is:

$$S_{n,n+1} = \frac{1}{\kappa^{(n)} + N_{x,n+1}(\kappa^{(n+1)} - \kappa^{(n)})} \begin{pmatrix} (\kappa^{(n+1)} - \kappa^{(n)}) V_{n+1} & \kappa^{(n+1)} \\ \kappa^{(n)} & N_{x,n+1}(1 - N_{x,n+1})(\kappa^{(n+1)} - \kappa^{(n)}) V_{n+1}^{-1} \end{pmatrix} \tag{73}$$

where $V_n = a_n b_n c_n$.

Dipole moment: When we are only interested in the scattered far field, the inclusion can be replaced by an equivalent dipole. When r is large in front of the lengths of the axes of the ellipsoid ($r \gg a, b, c$), ξ is of the order of r^2:

$$\xi \simeq r^2 \tag{74}$$

and the dipolar term is

$$\frac{B^{(o)}}{2} \int_{\xi}^{\infty} \frac{d\sigma}{(\sigma + a^2)R(\sigma)} \simeq \frac{B^{(o)}}{2} \int_{r^2}^{\infty} \frac{d\sigma}{\sigma^{5/2}} = \frac{B^{(o)}}{3r^3} \tag{75}$$

The component $\mathcal{P}_{elli,a}$ of dipole moment along the direction of the fluid flow is obtained by identification of the terms in r^{-2} in relations (46) and (75), namely:

$$\mathcal{P}_{elli,a} = \frac{4\pi}{3} U_{\infty} B^{(o)}. \tag{76}$$

2.3.4 Inclusion with continuously variable permeability

For ellipsoidal inclusion, we assume that mobility depends only on the variable ξ. From (33), the problem comes down to finding of the differential equation of the function $F(\xi)$. Eq. 33 is then:

$$\frac{d}{d\xi}\left(\kappa(\xi)(\xi + a^2)R(\xi)\frac{dF}{d\xi}\right) + \frac{d\kappa(\xi)}{d\xi}\frac{R(\xi)}{2}F(\xi) - \frac{d\kappa(\xi)}{d\xi}\frac{R(\xi)}{2}E = 0. \tag{77}$$

It is easy to verify that when κ is constant, we find the case of the homogeneous inclusion, and that if we put $a = b = c$, then we find the result of spherical inclusion.

2.4 Anisotropic defects

Often the defects occurring in porous media are anisotropic, i.e., some of their physical parameters like permeability are no longer scalar quantities but are tensors. For an anisotropic porous medium, assuming the Einstein convention, the Darcy's law is

$$v_i = -\frac{k_{ij}}{\eta}\partial_j p \quad \text{where} \quad \partial_j p = \frac{\partial p}{\partial x_j}. \tag{78}$$

The permeability is then defined by nine components k_{ij}, i.e., it has different values in different directions of the space. Liakopoulos [13] had shown that the permeability is a symmetric tensor of second rank. This leads to great simplifications for the study of such porous media. If in isotropic media the fluid velocity is aligned with the hydraulic gradient, in anisotropic media, this is true only along the principal directions of the tensor. It is therefore not surprising that the flow movement of the fluid is seriously disturbed by this type of defects.

In a 3D space, the permeability tensor has three principal directions perpendicular to each other and for which the permeability corresponds to the tensor eigenvalues. In the coordinates system defined by these directions, the permeability tensor is diagonal:

$$\mathbf{k} = \begin{pmatrix} k_1 & 0 & 0 \\ 0 & k_2 & 0 \\ 0 & 0 & k_3 \end{pmatrix}. \tag{79}$$

By its definition, mobility inherits properties of symmetry of permeability and is therefore a symmetric tensor such that $\kappa_{ij} = k_{ij}/\eta$.

The principal directions of the permeability tensors of the host medium and of the inclusion define coordinate systems which generally do not coincide. The system linked to the host environment is called the primary system, while that of the inclusion is called the secondary system.

In this part, the study of the effects of an anisotropic spherical inclusion in a porous medium explores three different configurations:

1. the host medium is isotropic and the defect is anisotropic,

2. the host medium is anisotropic and the defect is isotropic, and

3. the host medium and the defect are anisotropic.

2.4.1 Anisotropic defect in isotropic medium

In this configuration, the primary system is such that the incident pressure gradient is along the Oz axis and the secondary coordinate system is defined by the principal directions of the permeability tensor. Then, let (θ_0, φ_0) and (θ, φ) be the angular directions of the incident pressure gradient $\nabla p^{(o)}$ and of the observation vector $\mathbf{OM} = \mathbf{r}$ in the primary system. We note β as the angle between these two directions.

The external pressure verifies the classical Laplace equation, while the internal one is solution of the following equation:

$$\kappa_1^{(i)} \frac{\partial p^{(i)}}{\partial x^2} + \kappa_2^{(i)} \frac{\partial p^{(i)}}{\partial y^2} + \kappa_3^{(i)} \frac{\partial p^{(i)}}{\partial z^2} = 0. \tag{80}$$

This one is transformed into a Laplace equation as follows: at first, we dimensionalize the mobilities $\kappa_j^{(i)}$ by introducing the scalar quantity $\kappa'^{(i)}$. Eq. (80) then becomes

$$\kappa'^{(i)} \left(\kappa_{r,1}^{(i)} \frac{\partial p^{(i)}}{\partial x^2} + \kappa_{r,2}^{(i)} \frac{\partial p^{(i)}}{\partial y^2} + \kappa_{r,3}^{(i)} \frac{\partial p^{(i)}}{\partial z^2} \right) = 0, \tag{81}$$

where $\kappa_{r,j}^{(i)} = \kappa_j^{(i)} / \kappa'^{(i)}$. Using the linear transformation:

$$\frac{\partial}{\partial x_i'} = \sqrt{\kappa_{r,i}^{(i)}} \frac{\partial}{\partial x_i}, \tag{82}$$

Eq. (80) takes the form:

$$\kappa'^{(i)}\left(\frac{\partial^2 p^{(i)}}{\partial^2 x'}+\frac{\partial^2 p^{(i)}}{\partial^2 y'}+\frac{\partial^2 p^{(i)}}{\partial^2 z'}\right)=0. \tag{83}$$

The solutions $p^{(i)}$ and $p^{(o)}$, respectively, are:

$$p^{(i)}(r,\theta,\varphi)=\sum_{m,n} A^{(i)}_{m,n} r^n P^m_n(\cos\theta)\,cos\,(m\varphi)+\sum_{m,n} B^{(i)}_{m,n} r^n P^m_n(\cos\theta)\,sin\,(m\varphi), \tag{84}$$

$$\begin{aligned} p^{(o)}(r,\theta,\varphi) &= \sum_{m,n}\left(A^{(o)}_{m,n} r^n+\frac{B^{(o)}_{m,n}}{r^{n+1}}\right)P^m_n(\cos\theta)\,cos\,(m\varphi) \\ &+\sum_{m,n}\left(C^{(o)}_{m,n} r^n+\frac{D^{(o)}_{m,n}}{r^{n+1}}\right)P^m_n(\cos\theta)\,sin\,(m\varphi). \end{aligned} \tag{85}$$

In (84), only the finite terms at $r=0$ appear.

The amplitudes $A^{(o)}_{m,n}$, $B^{(o)}_{m,n}$, $C^{(o)}_{m,n}$, and $D^{(o)}_{m,n}$, are determined by the boundary conditions at $r=a$ and when $r\to\infty$.

When $r\to\infty$, $p^{(o)}$ is:

$$p^{(o)}\to E_\infty r\cos\beta, \quad \text{with} \quad E_\infty=-\frac{U_\infty}{\kappa^{(o)}}, \tag{86}$$

where β is the angle between vector $\mathbf{E}_\infty$ and the direction of the observer $\mathbf{OM}=\mathbf{r}$. If (θ,φ) resp. (θ_0,φ_0) are the angular coordinates of the observer (resp. of the incident pressure gradient), then

$$\cos\beta=\sin\theta\sin\theta_0\cos(\varphi-\varphi_0)+\cos\theta\cos\theta_0. \tag{87}$$

The expressions (86) and (87) show that, in the expansion (85), only nonzero terms are those for which $n=1$. Taking into account the relations $P^0_1(\cos\theta)=P_1(\cos\theta)=\cos\theta$ and $P^1_1(\cos\theta)=\sin\theta$, we obtain:

$$\begin{aligned} p^{(o)}(r,\theta,\varphi) &= A^{(o)}_{0,1} r\cos\theta+A^{(o)}_{1,1} r\sin\theta\cos\varphi+C^{(o)}_{1,1} r\sin\theta\sin\varphi \\ &+\sum_{m,n}\frac{B^{(o)}_{m,n}}{r^{n+1}}P^m_n(\cos\theta)\,cos\,(m\varphi)+\sum_{m,n}\frac{D^{(o)}_{m,n}}{r^{n+1}}P^m_n(\cos\theta)\,sin\,(m\varphi). \end{aligned} \tag{88}$$

By identification with (86) with help of (87), we find:

$$A^{(o)}_{0,1}=E_\infty\cos\theta_0, \tag{89}$$

$$A^{(o)}_{1,1}=E_\infty\sin\theta_0\cos\varphi_0, \tag{90}$$

$$C^{(o)}_{1,1}=E_\infty\sin\theta_0\sin\varphi_0. \tag{91}$$

The boundary conditions at $r=a$ (continuity of the stress component τ_{rr} and conservation of the fluid flow through the inclusion surface) are:

$$p^{(o)}\big|_{r=a}=p^{(i)}\big|_{r=a} \tag{92}$$

$$\kappa^{(o)} \frac{\partial p^{(o)}}{\partial r}\bigg|_{r=a} = \left(\kappa_{11}^{(i)} \frac{\partial p^{(i)}}{\partial r} + \kappa_{12}^{(i)} \frac{\partial p^{(i)}}{r\partial \theta} + \kappa_{13}^{(i)} \frac{\partial p^{(i)}}{r \sin\theta \partial \varphi} \right)\bigg|_{r=a}. \tag{93}$$

In these equations, $\kappa_{nm}^{(i)}$ are the components of the tensor $\kappa^{(i)}$ in the spherical coordinates given in Appendix D. These relations lead to the following expressions of the pressure:

$$p^{(i)}(r,\theta,\varphi) = \frac{3A\kappa_{(o)}}{2\kappa^{(o)} + \kappa_3^{(i)}} r\cos\theta + \frac{3B\kappa_{(o)}}{2\kappa^{(o)} + \kappa_1^{(i)}} r \sin\theta\cos\varphi + \frac{3D\kappa_{(o)}}{2\kappa^{(o)} + \kappa_2^{(i)}} r\sin\theta\sin\varphi, \tag{94}$$

$$p^{(o)}(r,\theta,\varphi) = Ar\cos\theta + Br\sin\theta\cos\varphi + Dr\sin\theta\sin\varphi + \frac{Aa^3}{r^2}\frac{\kappa^{(o)} - \kappa_3^{(i)}}{2\kappa^{(o)} + \kappa_3^{(i)}}\cos\theta$$
$$+ \frac{Ba^3}{r^2}\frac{\kappa^{(o)} - \kappa_1^{(i)}}{2\kappa^{(o)} + \kappa_1^{(i)}}\sin\theta\cos\varphi + \frac{Da^3}{r^2}\frac{\kappa^{(o)} - \kappa_2^{(i)}}{2\kappa^{(o)} + \kappa_2^{(i)}}\sin\theta\sin\varphi, \tag{95}$$

where

$$A = E_\infty \cos\theta_0, \quad B = E_\infty \sin\theta_0 \cos\varphi_0, \quad D = E_\infty \sin\theta_0 \sin\varphi_0. \tag{96}$$

The first three terms of the right-hand side of (95) are due to the pressure gradient applied to the porous medium. The last three terms are the pressure induced by the hydraulic dipoles directed along the principal directions of the anisotropic sphere.

When $\mathbf{E}_\infty$ is along the Oz axis and for $\varphi_0 = 0, \theta_0 = 0$ and $\kappa_1^{(i)} = \kappa_2^{(i)} = \kappa_3^{(i)} = \kappa^{(i)}$, we find the internal and external pressures of isotropic spherical inclusions (19) and (20).

Moreover, from the relations (94) and (95), it is possible to obtain the directions of the pressure gradient and of the velocity field inside the defect.

The inside fluid velocity results from (94); its components are given by $v_j^{(i)} = -\kappa_j^{(i)} \partial_j p^{(i)}$, from which we obtain:

$$\mathbf{v}^{(i)} = -\frac{3B\kappa_1^{(i)}}{2\kappa^{(o)} + \kappa_1}\mathbf{i} - \frac{3D\kappa_2^{(i)}}{2\kappa^{(o)} + \kappa_2}\mathbf{j} - \frac{3A\kappa_3^{(i)}}{2\kappa^{(o)} + \kappa_3}\mathbf{k}. \tag{97}$$

This is the generalization to the 3D case of the result obtained for the spherical inclusion when the pressure gradient is along the axis Ox (19).

The inner product of $\mathbf{v}^{(i)}$ and of the incident field $\mathbf{U}_\infty$, gives the angle γ whose the internal fluid velocity is deflected by the anisotropy of the inclusion:

$$\cos\gamma = \frac{\mathbf{U}_\infty \cdot \mathbf{v}^{(i)}}{\|\mathbf{U}_\infty\| \|\mathbf{v}^{(i)}\|}, \tag{98}$$

$$= \frac{\frac{\sin^2\theta_0 \sin^2\varphi_0 \kappa_1^{(i)}}{2\kappa^{(e)} + \kappa_2^{(i)}} + \frac{\sin^2\theta_0 \cos^2\varphi_0 \kappa_2^{(i)}}{2\kappa^{(e)} + \kappa_1^{(i)}} + \frac{\cos^2\theta_0 \kappa_3^{(i)}}{2\kappa^{(e)} + \kappa_2^{(i)}}}{\sqrt{\frac{\sin^2\theta_0 \sin^2\varphi_0 \kappa_1^{(i)}}{\left(2\kappa^{(e)} + \kappa_2^{(i)}\right)^2} + \frac{\sin^2\theta_0 \cos^2\varphi_0 \kappa_2^{(i)}}{\left(2\kappa^{(e)} + \kappa_1^{(i)}\right)^2} + \frac{\cos^2\theta_0 \kappa_3^{(i)}}{\left(2\kappa^{(e)} + \kappa_2^{(i)}\right)^2}}}. \tag{99}$$

2.4.2 Isotropic defect in anisotropic porous medium

Consider an isotropic sphere of radius r whose mobility is $\kappa^{(i)}$ which is included in an anisotropic host medium with its own mobility $\boldsymbol{\kappa}^{(o)}$. The incompressibility of the saturating fluid imposes that the outside pressure is the solution of the equation:

$$\partial_i \left(\kappa_{ij}^{(o)} \partial_j p \right) = 0. \tag{100}$$

In the system of Cartesian coordinate defined by the principal directions of the tensor $\boldsymbol{\kappa}^{(o)}$, this equation is written as:

$$\kappa'^{(o)} \left(\frac{\kappa_x^{(o)}}{\kappa'^{(o)}} \frac{\partial^2 p^{(o)}}{\partial x^2} + \frac{\kappa_y^{(o)}}{\kappa'^{(o)}} \frac{\partial^2 p^{(o)}}{\partial y^2} + \frac{\kappa_z^{(o)}}{\kappa'^{(o)}} \frac{\partial^2 p^{(o)}}{\partial z^2} \right) = 0, \tag{101}$$

where $\kappa_j^{(o)}$ are the eigenvalues of the mobility tensor and $\kappa'^{(o)}$ is an arbitrary scalar such that the ratio $\kappa_{r,j}^{(o)} = \kappa_j^{(o)} / \kappa'^{(o)}$ is a dimensionless quantity. Using the linear transformation of coordinates:

$$x' = \frac{x}{\sqrt{\kappa_{r,x}^{(o)}}}, \quad y' = \frac{y}{\sqrt{\kappa_{r,y}^{(o)}}}, \quad z' = \frac{z}{\sqrt{\kappa_{r,z}^{(o)}}}, \tag{102}$$

Eq. (101) becomes a Laplace equation. Correspondingly, the sphere is transformed into an ellipsoid with the semiaxes $a_x = r/\sqrt{\kappa_{r,x}^{(o)}}$, $a_y = r/\sqrt{\kappa_{r,y}^{(o)}}$, and $a_z = r/\sqrt{\kappa_{r,z}^{(o)}}$. Since the principal directions of the inside permeability $\kappa^{(i)}$ coincide with the axes of the ellipsoid, for each direction j, we find, for each of the components of the pressure gradient, the result of the ellipsoidal inclusion (40). The internal pressure gradient is then:

$$\partial_j p^{(i)} = \frac{\kappa'^{(o)}}{\kappa'^{(o)} + N_j \left(\kappa^{(i)} / \kappa_{rj}^{(o)} - \kappa'^{(o)} \right)} \partial_j p^{(o)}, \tag{103}$$

or

$$\partial_j p^{(i)} = \frac{\kappa_j^{(o)}}{\kappa_j^{(o)} + N_j \left(\kappa^{(i)} - \kappa_j^{(o)} \right)} \partial_j p^{(o)}. \tag{104}$$

In this equation, the depolarization factor N_j is

$$N_j = \frac{a_x a_y a_z}{2} \int_0^\infty \frac{ds}{\left(s + a_j^2 \right) \sqrt{\left(s + a_x^2 \right) \left(s + a_y^2 \right) \left(s + a_z^2 \right)}} \quad \text{for} \quad j = x, y, z. \tag{105}$$

Thus, the anisotropy induced in the sphere by the change of variables appears through the depolarization factor N_j.

2.4.3 Anisotropic defect in anisotropic porous medium

We assume now that the host medium and the defect have their own anisotropic microstructure with the mobilities tensors $\kappa_{ij}^{(o)}$ and $\kappa_{ij}^{(e)}$. The velocity of the fluid flowing in each part of the porous medium is given by equations:

$$v_i^{(o)} = -\kappa_{ij}^{(o)} \nabla_j p^{(o)}, \qquad v_i^{(i)} = -\kappa_{ij}^{(i)} \nabla_j p^{(i)}. \tag{106}$$

Without restricting the generality of the problem, the first relation of (106) can be written as:

$$v_i^{(o)} = -\kappa_i^{(o)} \frac{\partial p^{(o)}}{\partial x_i}, \tag{107}$$

where $\kappa_j^{(o)}, j = 1, 2, 3$, are the eigenvalues of the tensor $\boldsymbol{\kappa}^{(o)}$ and $v_i^{(o)}$ and $\frac{\partial p^{(o)}}{\partial x_i}$ are the components of the velocity and of the pressure gradient along the principal directions of this tensor.

The incompressibility of the fluid implies the condition:

$$\nabla_i v_i^{(o)} = 0, \tag{108}$$

or

$$\kappa_1^{(o)} \frac{\partial^2 p^{(o)}}{\partial x_1^2} + \kappa_2^{(o)} \frac{\partial^2 p^{(o)}}{\partial x_2^2} + \kappa_3^{(o)} \frac{\partial^2 p^{(o)}}{\partial x_3^2} = 0. \tag{109}$$

To transform this equation into a Laplace equation, we proceed as before by using the change of variables

$$\frac{\partial}{\partial x_i'} = \sqrt{\kappa_{r,i}^{(o)}} \frac{\partial}{\partial x_i}. \tag{110}$$

Then, the external environment becomes an isotropic medium and the outside pressure is the solution of the Laplace equation:

$$\frac{\partial p^{(o)}}{\partial x_1'^2} + \frac{\partial p^{(o)}}{\partial x_2'^2} + \frac{\partial^2 p^{(o)}}{\partial x_3'^2} = 0. \tag{111}$$

The new x_i' variables constitute a new coordinate system. The host medium is transformed into an isotropic medium, while the inclusion medium becomes anisotropic. In the new coordinate system, the pressure gradient is transformed according to:

$$\nabla' p = \sqrt{\kappa_{r,j}^{(o)}} \nabla p, \tag{112}$$

while the components of the position vector become:

$$r_i' = \left(\kappa_{r,i}^{(o)}\right)^{-1/2} r_i. \tag{113}$$

Using the Darcy's law and (112), the incompressibility of the fluid inside the inclusion

$$\nabla_i\left(\kappa_{ij}^{(i)}\nabla_j p^{(i)}\right) = 0 \tag{114}$$

implies the new equation:

$$\nabla_i'\left[\left(\kappa_i^{(o)}\right)^{-1/2}\kappa_{ij}^{(i)}\left(\kappa_j^{(o)}\right)^{-1/2}\nabla_j' p^{(i)}\right] = 0. \tag{115}$$

In the new coordinates system, the mobility in the inclusion defined by the equation:

$$\nabla_i'\left(\kappa_{ij}^{(i)'}\nabla_i' p_j^{(i)}\right) = 0 \tag{116}$$

is such that:

$$\kappa_{ij}^{(i)'} = \left(\kappa_{r,i}^{(o)}\right)^{-1/2}\kappa_{ij}^{(i)}\left(\kappa_{r,j}^{(o)}\right)^{-1/2}. \tag{117}$$

In the coordinates x_i', the semiaxes of the inclusion can be calculated from the equation of the ellipsoidal inclusion surface written in matrix form as R^tAR, where R is the position vector of a point of this surface ($R^t = (x, y, z)$) and A is the diagonal matrix whose entries are the lengths of the half-axes:

$$A = \begin{pmatrix} a_1 & 0 & 0 \\ 0 & a_2 & 0 \\ 0 & 0 & a_3 \end{pmatrix}. \tag{118}$$

Then, the linear transformation (113) changes A into A':

$$A' = \begin{pmatrix} a_1' & 0 & 0 \\ 0 & a_2' & 0 \\ 0 & 0 & a_3' \end{pmatrix}, \tag{119}$$

with

$$a_i'^2 == \left(\kappa_{r,i}^{(o)}\right)^{-1/2} a_i^2 \left(\kappa_{r,i}^{(o)}\right)^{-1/2}. \tag{120}$$

Thus, the operation that transforms the anisotropic host medium into an isotropic one transforms the ellipsoidal inclusion with the semiaxes (a_1, a_2, a_3) into another one with the new semiaxes (a_1', a_2', a_3') and the new mobility $\kappa_{ij}^{(i)'}$ given respectively by (120) and.

We recover the previous case where the outer medium is isotropic and the inner medium is anisotropic. So, in accordance with (104):

$$\partial_j p^{(i)} = \frac{\kappa'^{(o)}}{\kappa'^{(o)} + N_j'\left(\kappa^{(i)}/\kappa_{rj}^{(o)} - \kappa'^{(o)}\right)}\partial_j p^{(o)}, \tag{121}$$

where the depolarization factors of the new inclusion are given by:

$$N'_i = \frac{detA'}{2}\int_0^\infty \frac{d\sigma}{(a'_i+\sigma)\sqrt{det\left(A'^2+\sigma I\right)}}. \tag{122}$$

2.5 Hydraulic polarisability

As mentioned above, the reaction of a saturated porous inclusion subject to a pressure gradient is to induce a hydraulic dipole whose dipole moment is $\mathcal{P}$. This dipole results from the appearance of pressure discontinuities at the inclusion-host interface. They have different signs depending on whether the flow is incoming or outgoing, but have the same absolute value. They are the hydraulic analogues of electrostatic charges induced by an electric field in a dielectric medium. The resulting hydraulic polarization is only nonzero if the contrast between the mobility of the host environment and that of inclusion is itself nonzero.

$$\mathcal{P} = \int_\Omega \left(\kappa^{(i)}(\mathbf{r}) - \kappa^{(o)}\right)\kappa^{(i)}\nabla p^{(i)}(\mathbf{r})dV. \tag{123}$$

For spherical or ellipsoidal inclusions and for low filtration rates, we have seen that the internal pressure gradient is proportional to the incident one. For this type of inclusions, the dipole moment is written as:

$$\mathcal{P} = \alpha\mathbf{v}^{(o)}, \tag{124}$$

where the value of the susceptibility α measures the ability of the inclusion to induce a dipole under the action of a pressure gradient. α can be seen as the "hydraulic polarisability" of the defect. For a spherical defect of volume V, we have:

$$\mathcal{P} = \alpha(-\mathbf{U}_\infty) \tag{125}$$

with

$$\alpha = 3V\kappa^{(i)}\left(\frac{\frac{k^{(i)}}{k^{(o)}}-1}{2+\frac{k^{(i)}}{k^{(o)}}}\right). \tag{126}$$

For an ellipsoidal inclusion, hydraulic polarisability is not a scalar since the response of the inclusion is a function of the direction of pressure incidence, but a second rank tensor whose eigenvalues are the susceptibilities along the three axes of the ellipsoid:

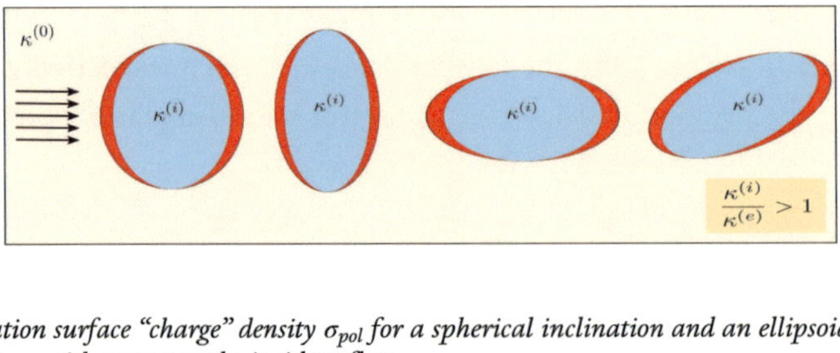

Figure 11.
Sketch of polarization surface "charge" density σ_{pol} for a spherical inclination and an ellipsoidal inclusion with different orientations with respect to the incident flow.

$$\alpha_i = V\kappa^{(i)} \frac{\kappa^{(i)} - \kappa^{(o)}}{\kappa^{(o)} + N_i\left(\kappa^{(i)} - \kappa^{(o)}\right)}, \quad i = x, y, z. \tag{127}$$

Figure 11 represents the surface hydraulic charges induced by the hydraulic polarization on a spherical inclusion and on ellipsoidal inclusions with different orientations with respect to the direction of the incident flux. The red areas represent the surface "hydraulic charge" density σ_{pol}. Its expression depends on the direction of the incident pressure gradient and is written as the sum of the contributions of the dipoles along the three axes of the ellipsoid.

When the permeability of the inclusion is stratified, the dipole moment is given by the dipolar term ($B^{(o)}$) of the external pressure field obtained by the transfer matrix method or by the scattering matrix method.

3. Tortuosity induced by defects

In this section, we determine the hydraulic effects of defects on the permeability of porous media. As mentioned above, the shape of the defects is one of the most important factors for the modification of the current lines of the seepage rate in the whole porous medium and thus contributes to its acoustic properties.

3.1 Homogenization: generalities

Experiments show that a nonhomogeneous medium subject to excitation behaves in the same way as its different components, but with different parameter values. The homogenization of an inhomogeneous porous medium consists in replacing it with an effective homogeneous medium with the permeability k_{eff}. This operation is only possible at a fairly large observation scale. Determining the value of the effective permeability from the mobility values of the structure components and their relative positions is not a simple averaging operation. The calculation of the global mobility of a mixture of porous inclusions immersed in a homogeneous medium is a topic widely addressed in many research fields such as hydrology, oil recovery, chemical industry, etc. As a consequence, a considerable number of works deal with this problem based on various methods: renormalization theory, variational methods, T-Matrix method, field theory methods, nonperturbative approach based on Feynman path integral. To quote some of authors, we can refer to the works of Prakash and Raja-Sekhar [14], King [15, 16], Drummond and Horgan [17], Dzhabrailov and Meilanov [18], Teodorovich [19, 20], Stepanyants and Teodorovich [21], and Hristopulos and Christakos [22].

In the case of media subject to a variable field action, homogenization requires defining a length below which it is no longer relevant. For example, for a periodic field, acting on a medium whose average distance between inhomogeneities is a, this length is the wavelength λ if $\lambda/a \gg 1$. In our case, effective mobility being essentially a low-frequency concept, this remark justifies that the effective mobility should then be calculated from a steady filtration velocity.

3.2 Effective mobility

Darcy's law is often used as the definition of the mobility of a porous medium, and the easiest way to introduce the effective mobility κ_{eff} is to use it as follows:

$$< \mathbf{v} > = \kappa_{eff} < \mathbf{E} >, \tag{128}$$

where $\mathbf{E} = -\nabla p$ and $< \cdot >$ is the averaging operation. The mean values of the filtration rate and the pressure gradient are given by:

$$< \mathbf{v} > = f < \kappa^{(i)} \mathbf{E}^{(i)} > + (1-f) < \kappa^{(o)} \mathbf{E}^{(o)} >, \tag{129}$$

$$< \mathbf{E} > = f < \mathbf{E}^{(i)} > + (1-f) < \mathbf{E}^{(o)} >, \tag{130}$$

where f is the volumic fraction of the defect. Putting $\mathbf{E}^{(i)} = \mathcal{A}\mathbf{E}^{(o)}$ we show that:

$$\kappa_{eff} = \frac{f \mathcal{A} \kappa^{(i)} + (1-f)\kappa^{(o)}}{f\mathcal{A} + (1-f)}. \tag{131}$$

For spherical defects, $\mathcal{A} = 3\kappa^{(o)} / (2\kappa^{(o)} + \kappa^{(i)}$, Eq. (131) leads to the result:

$$\kappa_{eff} = \kappa^{(o)} \frac{f \frac{3\kappa^{(i)}}{2\kappa^{(o)} + \kappa^{(i)}} + (1-f)}{f \frac{3\kappa^{(o)}}{2\kappa^{(o)} + \kappa^{(i)}} + (1-f)}. \tag{132}$$

When $f \to 0$, then $\kappa_{eff} \sim \kappa^{(o)}$, and when $f \to 1$, then $\kappa_{eff} \sim \kappa^{(i)}$. Finally when $f < < 1$, then

$$\kappa_{eff} \sim \kappa^{(o)} + 3f\kappa^{(o)} \frac{\kappa^{(i)} - \kappa^{(o)}}{\kappa^{(i)} + 2\kappa^{(o)}}. \tag{133}$$

For anisotropic inclusion, mobility is a second rank tensor defined by the relationship:

$$< v_i > = \kappa_{eff,ij} < E_j >. \tag{134}$$

As a result, for each main direction, we have:

$$< v_j > = \kappa_{eff,j} < E_j >, \quad j = x, y, z, \tag{135}$$

where $\kappa_{eff,j}$ are the eigenvalues of κ_{eff}. Taking into account that $E_j^{(i)} = \mathcal{A}_j E_j^{(o)}$, Eq. (135) shows that:

$$\kappa_{eff,j} = \frac{f \mathcal{A}_j \kappa^{(i)} + (1-f)\kappa^{(o)}}{f\mathcal{A}_j + (1-f)}, \tag{136}$$

with

$$\mathcal{A}_j = \frac{3\kappa^{(o)}}{2\kappa^{(o)} + \kappa_j^{(i)}}, \quad j = 1, 2, 3. \tag{137}$$

When the environment has several defects, the calculation of k_{eff} is more complicated because their mutual influence must be taken into account. The excitation pressure gradient $\mathbf{E}^e$ defined from the filtration rate is introduced by the equation:

$$\kappa^{(o)}\mathbf{E}^e = \kappa^{(o)}\mathbf{E} + L \cdot P. \tag{138}$$

In this relationship, L is an operator that takes into account the shape of the defect and its orientation with respect to the fluid flow, and P is due to the induced polarization in the inclusion. P defined by (123) is related to the dipole moment induced by the interaction between the fluid moving in the porous medium and the defect. When the medium contains n identical defects per unit volume, $P = np$, p being the dipole moment of each defect, and since p is proportional to the applied field ($p = \alpha\mathbf{E}^e$), we have $P = n\alpha\mathbf{E}^e$. For an ellipsoidal defect, L is reduced to depolarization factors, i.e., $L = N_k$, $k = x, y, z$, which takes into account the direction of fluid flow. In this case, the excitation field is:

$$E^e = \frac{E}{1 - n\alpha N_k/\kappa^{(o)}}, \tag{139}$$

leading to the following κ_{eff} expression:

$$\kappa_{eff} = \kappa^{(o)} + \frac{n\alpha}{1 - n\alpha N_k/\kappa^{(o)}}, \tag{140}$$

leading, for spherical defects, to expression:

$$\kappa_{eff} = \kappa^{(o)} + \frac{n\alpha}{1 - n\alpha/3\kappa^{(o)}}. \tag{141}$$

It is then possible to calculate the effective mobility of a set of ellipsoidal inclusions in different geometries (**Figure 12**):

- the ellipsoids are aligned with the direction x of the fluid flow:

$$\kappa_{eff} = \kappa^{(o)} + \frac{n\alpha_x}{1 - n\alpha_x N_x/\kappa^{(o)}}. \tag{142}$$

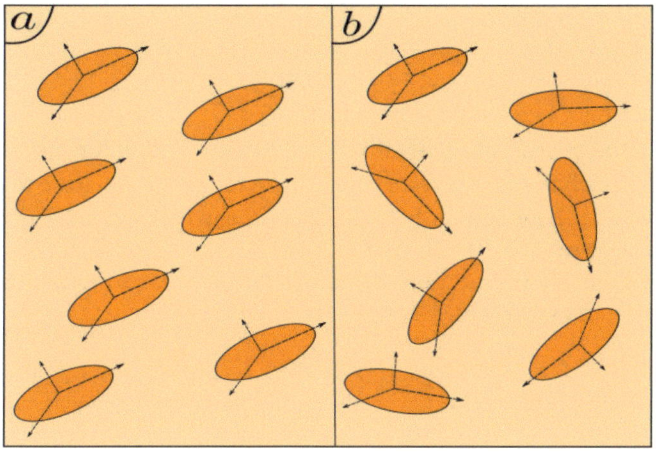

Figure 12.
Sets of ellipsoidal inclusions: (a) aligned and (b) randomly oriented.

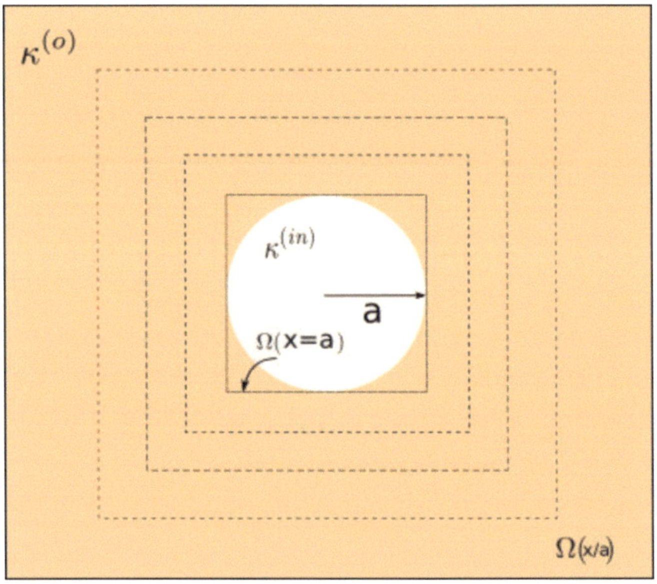

Figure 13.
Domains used in numerical simulations for the evaluation of $<v^{(o)2}>$.

- the ellipsoids are randomly oriented:

$$\kappa_{eff} = \kappa^{(o)} + \frac{1/3 \sum_{i=x,\,y,\,z} n\alpha_i}{1 - \sum_{i=x,\,y,\,z} n\alpha_i N_i / \kappa^{(o)}}. \tag{143}$$

3.3 Induced tortuosity

We restrict ourselves here to the calculation of the tortuosity induced by homogeneous spherical inclusions (**Figure 13**). Since the dipole moment is the essential element for this calculation element for this calculation, it is easy to generalize the results obtained with other types of inclusions: inhomogeneous spherical, ellipsoidal, etc.

The tortuosity induced by the presence of defects τ_d is defined by:

$$\tau_d = \frac{<v^{(o)2}>}{<v^{(o)}>^2} \tag{144}$$

where $v^{(o)}$ is the perturbation of filtration rate due to the defects. When the ratio k/a^2 where a is the characteristic size of the defects is small relative to the unity, it is legitimate to neglect the volume of the defects for the calculation of $<v^2>$, whereas it is taken into account for that of $<v>$.

With the pressure scattered field by the inclusions being limited to the dipolar terms, the expression of $<v^{(o)2}>$ is then:

$$<v^{(o)2}> = <v_r^{(o)2}> + <v_\theta^{(o)2}> \tag{145}$$

where

$$v_r^{(o)} = -\kappa^{(o)} \frac{\partial p^{(o)}}{\partial r}, \text{ and } v_\theta^{(o)} = -\kappa^{(o)} \frac{\partial p^{(o)}}{r\partial \theta}. \tag{146}$$

For ellipsoidal inclusion, the external pressure is:

$$p^{(o)}(r,\theta) = -\frac{U_\infty}{\kappa^{(o)}} r\cos\theta + \frac{\mathcal{P}_d}{4\pi}\frac{\cos\theta}{r^2} \tag{147}$$

$$= -\frac{U_\infty}{\kappa^{(o)}}\left(r - \frac{\alpha}{r^2}\right)\cos\theta \tag{148}$$

where $\mathcal{P}_d$ and α are, respectively, the dipol moment and the polarisability of the inclusion. By keeping only the terms greater than or equal to r^{-2}, one obtains:

$$v^{(o)2} \approx \frac{U_\infty}{\kappa^{(o)}}^2\left[1 - \frac{2\alpha}{r^3}\left(2\cos^2\theta - \sin^2\theta\right)\right] \tag{149}$$

The average value $<v^{(o)2}>$ is calculated by integration on the volume between two spheres of radius a (characteristic size of the defect) and R sufficiently large so that the dipolar effects are negligible. For a spherical inclusion, it results:

$$<v^{(o)2}> = U_\infty^2 + U_\infty^2\left(1 - \frac{a^3}{R^3}\right)\left(\frac{\frac{\kappa^{(i)}}{\kappa^{(o)}} - 1}{2 + \frac{\kappa^{(i)}}{\kappa^{(o)}}}\right)^2. \tag{150}$$

$<v^{(o)}>^2$ is calculated from the definition of effective mobility:

$$<v> = \kappa_{eff} <E> \tag{151}$$

where κ_{eff} and $<E>$ are given by (131) and (130). When $f << 1$, we have

$$\kappa_{eff} \approx \kappa^{(o)} + f\mathcal{A}\left(\kappa^{(i)} - \kappa^{(o)}\right) \quad \text{and} \quad <E> \approx \frac{U_\infty}{\kappa^{(o)}}(1 + f\mathcal{A}). \tag{152}$$

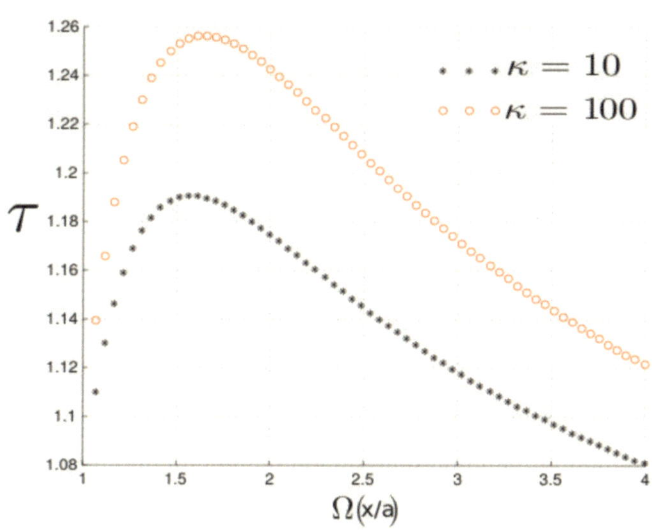

Figure 14.
Evaluation of the tortuosity induced as a function of the distance to the defect.

From these two relations, we obtain the expression of the induced tortuosity:

$$\tau_d = \left[1 + \left(1 - \frac{a^3}{R^3}\right)\left(\frac{\frac{\kappa^{(i)}}{\kappa^{(o)}} - 1}{2 + \frac{\kappa^{(i)}}{\kappa^{(o)}}}\right)^2\right]\left[1 - 2f\frac{3\frac{\kappa^{(i)}}{\kappa^{(o)}}}{2 + \frac{\kappa^{(i)}}{\kappa^{(o)}}}\right]. \tag{153}$$

Results of numerical simulations: The results of a numerical simulation for $\kappa^{(i)}/\kappa^{(o)} = 10$ and $\kappa^{(i)}/\kappa^{(o)} = 100$ are shown in **Figure 14**. The tortuosity value is calculated on square domains around the inclusion (**Figure 13**). Inside the inclusion, $\tau_b = 1$. As x increases, the tortuosity increases to reach its maximum value at $x = 1.7$ for $\kappa^{(i)}/\kappa^{(o)} = 10$ and $x = 1.6$ when $\kappa^{(i)}/\kappa^{(o)} = 10$. For larger values of x, it decreases toward 1 since, far from inclusion, the field lines again become parallel to the direction of the incident pressure gradient. This result confirms the behavior of the field lines of **Figure 4b**.

4. Conclusion

In this chapter, we studied the effect of defects on the circulation of the fluid saturating a porous medium. We have shown that the modification of the stream lines of the filtration velocities leads to a modification of the value of the tortuosity and thus on the local velocity of the waves susceptible to propagate in such media. The induced tortuosity was calculated from the pressure field scattered by the inclusions. The model used is based on the Darcy's law. in addition to being general, its major interest is to lead to a very practical mathematical expression of tortuosity

Acknowledgements

C. Depollier is supported by Russian Science Foundation grant number 14-49-00079.

A. Ellipsoidal coordinates

The ellipsoidal coordinates (ξ, η, ζ) are the solutions of the cubic equation:

$$\frac{x^2}{a^2 + u} + \frac{y^2}{b^2 + u} + \frac{z^2}{c^2 + u} = 1. \tag{154}$$

They are connected to the Cartesian coordinates (x, y, z) by the relations:

$$x^2 = \frac{(a^2 + \xi)(a^2 + \eta)(a^2 + \zeta)}{(b^2 - a^2)(c^2 - a^2)}, \tag{155}$$

$$y^2 = \frac{(b^2 + \xi)(b^2 + \eta)(b^2 + \zeta)}{(a^2 - b^2)(c^2 - b^2)}, \tag{156}$$

$$z^2 = \frac{(c^2+\xi)(c^2+\eta)(c^2+\zeta)}{(a^2-c^2)(b^2-c^2)}, \tag{157}$$

subject to the conditions $-\xi < c^2 < -\eta < b^2 < -\zeta < a^2$.
The scalar factors are the vector norms:

$$h_{q_i} = \| \frac{\partial \mathbf{r}}{\partial q_i} \| \quad \text{òu} \quad q_i = \xi, \eta, \zeta. \tag{158}$$

Their values are:

$$h_\xi = \frac{1}{2}\sqrt{\frac{(\eta-\xi)(\zeta-\xi)}{(a^2-\xi)(b^2-\xi)(c^2-\xi)}}, \tag{159}$$

$$h_\eta = \frac{1}{2}\sqrt{\frac{(\xi-\eta)(\zeta-\eta)}{(a^2-\eta)(b^2-\eta)(c^2-\eta)}}, \tag{160}$$

$$h_\zeta = \frac{1}{2}\sqrt{\frac{(\eta-\zeta)(\xi-\zeta)}{(a^2-\zeta)(b^2-\zeta)(c^2-\zeta)}}. \tag{161}$$

B. Depolarization factors

The depolarization factors are important quantities for the expression of solutions of the Laplace equation. They take into account the form of the domain in which this solution is sought and its orientation in relation to the excitation field. Their expression is:

$$N_k = \frac{abc}{3}\int_0^\infty \frac{d\sigma}{(\sigma+q_k^2)\sqrt{(\sigma+a^2)(\sigma+b^2)(\sigma+c^2)}} \tag{162}$$

where $k = x$ (resp. y, z) et $q_k = a$, (resp. b, c) and satisfy the relation:

$$N_x + N_y + N_z = 1. \tag{163}$$

C. Relations between two spherical coordinates systems

Consider the rectangular coordinate systems (x, y, z) and (x', y', z'). We are looking for the relations between the spherical coordinates (r, θ, φ) and (r', θ', φ') associated with each of them. From

$$x' = r' \sin\theta' \cos\varphi' \qquad x = r\sin\theta\cos\varphi, \tag{164}$$

$$y' = r' \sin\theta' \sin\varphi' \qquad y = r\sin\theta\sin\varphi, \tag{165}$$

$$z' = r' \cos\theta' \qquad z = r\sin\theta, \tag{166}$$

one deduces

$$\begin{aligned} r'^2 &= x'^2 + y'^2 + z'^2 \\ &= \frac{x^2}{\kappa_1} + \frac{y^2}{\kappa_2} + \frac{z^2}{\kappa_3} \\ &= r^2\left(\frac{\sin^2\theta\cos^2\varphi}{\kappa_1} + \frac{\sin^2\theta\sin^2\varphi}{\kappa_2} + \frac{\cos^2\theta}{\kappa_3}\right) \end{aligned}$$

or

$$r' = r\Delta \quad \text{où} \quad \Delta = \sqrt{\frac{\sin^2\theta\cos^2\varphi}{\kappa_1} + \frac{\sin^2\theta\sin^2\varphi}{\kappa_2} + \frac{\cos^2\theta}{\kappa_3}}. \tag{167}$$

From (164) and (165), one has:

$$\begin{aligned} x' &= \frac{x}{\kappa_1} \Rightarrow r'\sin\theta'\cos\varphi' &= \frac{r}{\kappa_1}\sin\theta\cos\varphi \\ y' &= \frac{y}{\kappa_2} \Rightarrow r'\sin\theta'\sin\varphi' &= \frac{r}{\kappa_2}\sin\theta\sin\varphi. \end{aligned}$$

By eliminating φ, one finds:

$$\sin\theta' = \sin\theta\frac{\delta}{\Delta} \quad \text{avec} \quad \delta = \sqrt{\frac{\cos^2\varphi}{\kappa_1} + \frac{\sin^2\varphi}{\kappa_2}}. \tag{168}$$

In the same way, from (166), one can establish

$$\cos\theta' = \frac{1}{\Delta}\frac{\cos\theta}{\sqrt{\kappa_3}}. \tag{169}$$

Similar relationships between angles φ and φ' are deduced from (164) and (165):

$$\sin\varphi' = \frac{1}{\delta}\frac{\sin\varphi}{\sqrt{\kappa_2}}, \tag{170}$$

$$\cos\varphi' = \frac{1}{\delta}\frac{\cos\varphi}{\sqrt{\kappa_1}}. \tag{171}$$

D. Tensor

Let $\boldsymbol{\kappa}$ be a tensor of rank 2. We denote by $\boldsymbol{\kappa}_r$, its expression in the system of rectangular coordinates defined by its principal directions, and $\boldsymbol{\kappa}_s$, its expression in the corresponding spherical coordinates system. So we have

$$\boldsymbol{\kappa}_r = \begin{pmatrix} \kappa_1 & 0 & 0 \\ 0 & \kappa_2 & 0 \\ 0 & 0 & \kappa_3 \end{pmatrix} \qquad \boldsymbol{\kappa}_s = \begin{pmatrix} \kappa_{11} & \kappa_{12} & \kappa_{13} \\ \kappa_{21} & \kappa_{22} & \kappa_{23} \\ \kappa_{31} & \kappa_{32} & \kappa_{33} \end{pmatrix} \tag{172}$$

with

$$\kappa_{11} = \kappa_1 \sin^2\theta \cos^2\varphi + \kappa_2 \sin^2\theta \sin^2\varphi + \kappa_3 \cos^2\theta, \tag{173}$$

$$\kappa_{12} = \kappa_1 \cos\theta \sin\theta \cos^2\varphi + \kappa_2 \cos\theta \sin\theta \sin^2\varphi - \kappa_3 \cos\theta \sin\theta, \tag{174}$$

$$\kappa_{13} = (\kappa_2 - \kappa_1) \sin\theta \cos\varphi \sin\varphi, \tag{175}$$

$$\kappa_{22} = \kappa_1 \cos^2\theta \cos^2\varphi + \kappa_2 \cos^2\theta \sin^2\varphi + \kappa_3 \sin^2\theta, \tag{176}$$

$$\kappa_{23} = (\kappa_2 - \kappa_1) \cos\theta \cos\varphi \sin\varphi, \tag{177}$$

$$\kappa_{33} = \kappa_1 \sin^2\varphi + \kappa_2 \cos^2\varphi, \tag{178}$$

$$\kappa_{21} = \kappa_{12}, \tag{179}$$

$$\kappa_{31} = \kappa_{13}, \tag{180}$$

$$\kappa_{32} = \kappa_{23}. \tag{181}$$

Or, alternatively in the matrix form:

$$\boldsymbol{k} = \kappa_1 \mathbf{I} + (\kappa_2 - \kappa_1)\mathbf{A} + (\kappa_3 - \kappa_1)\mathbf{B}, \tag{182}$$

where $\mathbf{I}$ is the unit matrix 3×3 and $\mathbf{A}$ and $\mathbf{B}$ are given by:

$$\mathbf{A} = \begin{pmatrix} \sin^2\theta sin^2\varphi & \cos\theta \sin\theta \sin^2\varphi & \sin\theta \cos\varphi \sin\varphi \\ \cos\theta \sin\theta \sin^2\varphi & \cos^2\theta \sin^2\varphi & \cos\theta \cos\varphi \sin\varphi \\ \sin\theta \cos\varphi \sin\varphi & \cos\theta \cos\varphi \sin\varphi & \cos^2\theta \end{pmatrix}, \tag{183}$$

$$\mathbf{B} = \begin{pmatrix} \cos^2\theta & -\cos\theta \sin\theta & 0 \\ -\cos\theta \sin\theta & \sin^2\theta & 0 \\ 0 & 0 & 0 \end{pmatrix}. \tag{184}$$

Author details

Fatma Graja[1] and Claude Depollier[1,2]*

1 LUNAM Université du Maine, UMR CNRS 6613 Laboratoire d'Acoustique de l'Université du Maine, Le Mans, France

2 MPEI, Moscow, Russia

*Address all correspondence to: claude.depollier@univ-lemans.fr

References

[1] Ghanbarian B, Hunt AG, Ewing RE, Sahimi M. Tortuosity in porous media: A critical review. Soil Science Society of America Journal. 2013;77:1461-1477

[2] Carman PC. Fluid flow through granular beds. Transactions Institute of Chemical Engineers. 1937;**15**:150-166

[3] Darcy HPG. Les fontaines publiques de la ville de Dijon. Paris: Dalmont; 1856

[4] Saomotoa H, Katagiri J. Direct comparison of hydraulic tortuosity and electric tortuosity based on finite element analysis. Theoretical and Applied Mechanics Letters. 2015;**5**: 177-180

[5] Johnson DL. Frontiers in Physical Acoustics. In: Proceedings of the Enrico Fermi Summer School, Course XCII. New York: Elsevier; 1984

[6] Lafarge D, Allard JF, Brouard B, Verhaegen C, Lauriks W. Characteristic dimensions and prediction at high frequencies of the surface impedance of porous layers. The Journal of the Acoustical Society of America. 1993;**93**: 2474-2478

[7] Brinkman HC. A calculation of the viscous force exerted by a flowing fluid on a dense swarm of particles. Applied Science Research. 1947;**A1**:27

[8] Schiffer M, Szegö G. Virtual mass and polarization. Transactions of the American Mathematical Society. 1949; **67**:130-205

[9] Stoner EC. The demagnetizing factors for ellipsoids. Philosophical Magazine, Series 7. 1945;**36**(263): 803-821

[10] Osborn JA. Demagnetizing factors of the general ellipsoid. Physical Review. 1945;**67**(1112):351-357

[11] Landau L, Lifchitz E. Electodynamique des milieux continus. Moscou: MIR; 1969

[12] Mistri F, Wang AP. The star-product and its algebraic properties. Journal of the Franklin Institute. 1986; **321**(1):21-38

[13] Liakopoulos AC. Darcy's coefficient of permeability as symmetric tensor of second rank. Hydrological Sciences Journal. 1965;**10**:41-48

[14] Prakash J, Raja Sekhar GP. Arbitrary oscillatory Stokes flow past a porous sphere using Brinkman model. Meccanica. 2012;**47**:1079-1095

[15] King PR. The use oof field theoretic methods for the study of flow in a heterogeneous porous medium. Journal of Physics A: Mathematical and General. 1987;**20**:3935-3947

[16] King PR. The use of renormalization for calculating effective permeability. Transport in Porous Media. 1989;**4**:37-58

[17] Drummond IT, Horgan RR. The effective permeability of a random medium. Journal of Physics A: Mathematical and General. 1987;**20**: 4661-4672

[18] Dzhabrailov VV, Meilanov RP. Filtration in a porous medium with a fluctuating permeability. Journal of Engineering Physics and Thermophysics. 1996;**69**:188-192

[19] Teodorovich EV. Calculation of the effective permeability of a randomly inhomogeneous porous medium. Journal of Experimental and Theoretical Physics. 1997;**85**:173-178

[20] Teodorovich EV. The effective "conductivity of a randomly inhomogeneous medium". Journal of

Applied Mathematics and Mechanics. 2000;**64**:951-957

[21] Stepanyants YA, Teodorovich EV. Effective hydraulic conductivity of a randomly heterogeneous porous medium. Water Resources Research. 2003;**39**:SHB 12-1-SHB 12-9

[22] Hristopulos DT, Christakos G. Variational calculation of the effective fluid permeability of heterogeneous media. Physical Review E. 1997;**55**: 7288-7298

Chapter 5

Acoustics and Biological Structures

Mariana Alves-Pereira, Bruce Rapley, Huub Bakker and Rachel Summers

Abstract

Within the context of noise-induced health effects, the impact of airborne acoustical phenomena on biological tissues, particularly within the lower frequency ranges, is very poorly understood. Although the human body is a viscoelastic-composite material, it is generally modeled as Hooke elastic. This implies that acoustical coupling is considered to be nonexistent at acoustical frequencies outside of the human auditory threshold. Researching the acoustical properties of mammalian tissue raises many problems. When tissue samples are investigated as to their pure mechanical properties, stimuli are not usually in the form of airborne pressure waves. Moreover, since the response of biological tissue is dependent on frequency, amplitude, and time profile, precision laboratory equipment and relevant physiological endpoints are mandatory requirements that are oftentimes difficult to achieve. Drawing upon the viscoelastic nature of biological tissue and the tensegrity model of cellular architecture, this chapter will visit what is known to date on the biological response to a variety of different acoustic stimuli at very low frequencies.

Keywords: infrasound, low frequency noise, health, cellular biology, tissue morphology

1. Introduction

Airborne pressure waves are ubiquitous in all human environments and have played vital roles in the survival, evolution, and development of the human species. Under certain conditions, airborne pressure waves can be perceived as "sound" by the human auditory system. Under other conditions, they may be perceived as a whole-body or partial-body vibration. Some airborne pressure waves are not consciously perceived at all. As human societies developed and became more technological, airborne pressure waves emanating from human-made devices became ubiquitous and "noise" became a more serious issue. By the late nineteenth century, noise and health studies began to flourish. In the early twentieth century, the telephone and growing industrialization led to more in-depth studies of the human hearing function. In 2011, a WHO document on the burden of diseases reflected the seriousness of the ongoing "noise problem" [1].

The only airborne pressure waves considered of consequence for human health were those that could be *heard*, i.e., "what you can't hear can't hurt you" (**Figure 1**). This notion justified the development of acoustic measuring devices and methodologies that concentrated solely on the audible portion of the acoustical spectrum.

Within the audible segment (20–20,000 Hz), human auditory acuity is not evenly distributed, and is more sensitive within the 800–7000 Hz range than it is to air-borne acoustic events occurring below 500 Hz or above 15,000 Hz. Thus, early on, scientists understood that in order to protect human hearing function and speech intelligibility, the entire audible segment need not be considered, but rather, only the frequencies at which the acuity was highest: 800–7000 Hz range. The development of the A-frequency weighting and the resulting deciBel-A (dBA) metric allowed acousticians and health professionals to assess acoustical environments simulating this variability of human auditory acuity.

Figure 2 shows the frequency response curve for the dBA metric, clearly following the human auditory response to airborne acoustic pressure waves.

While the dBA metric proved to be key for the protection of hearing and speech intelligibility, it was insufficient for the assessment of airborne pressure waves occurring outside of the 800–7000 Hz range. **Figure 3** emphasizes the 800–7000 Hz range within the dBA metric, and **Figure 4** shows its application at 10 Hz. The dBA metric is, therefore, unsuited for evaluating airborne pressure waves occurring at frequencies below 800 Hz. Health effects that may be developing due to exposures

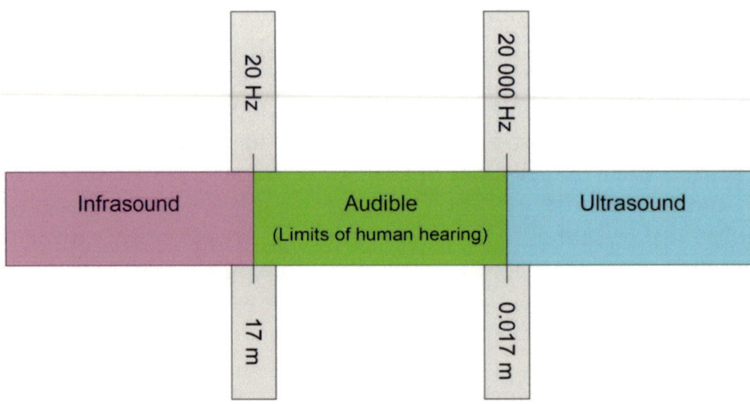

Figure 1.
Acoustical spectrum showing the classical three segments (infrasound, audible, and ultrasound) with the frequency and wavelength indicated at the cutoff of each segment.

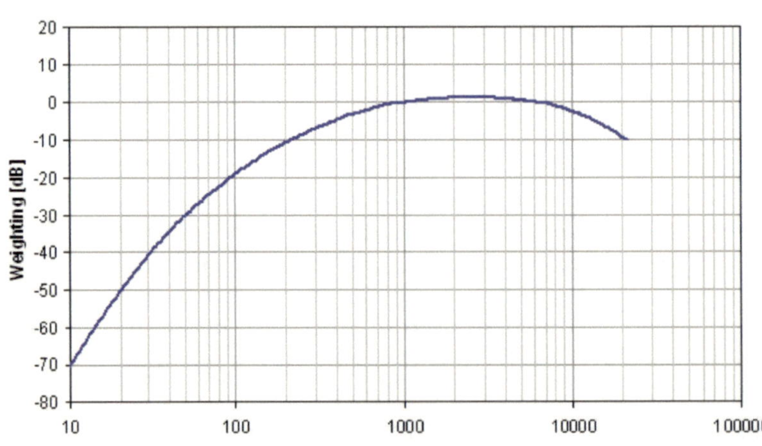

Figure 2.
Frequency response curve for the deciBel-A metric (dBA) commonly used in noise-related legislation [2].

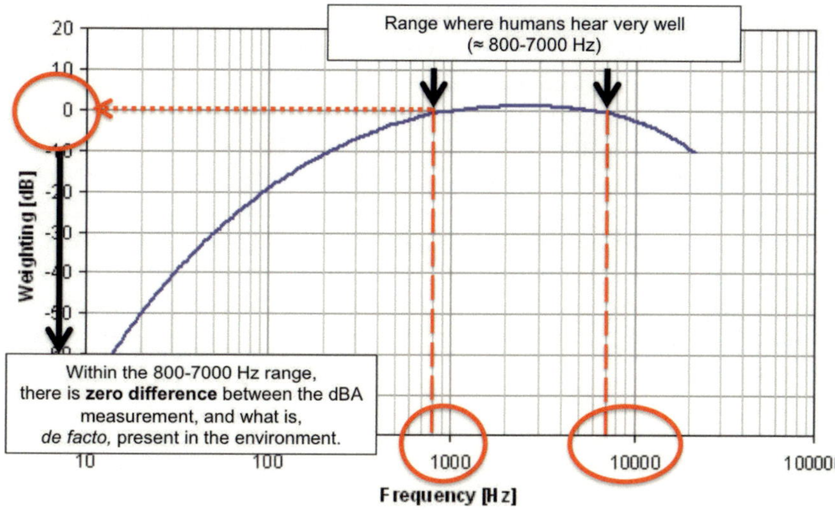

Figure 3.
Frequency response curve for the dBA metric applied to the range of highest human auditory acuity. Within this frequency range, the dBA measurement will accurately reflect the airborne acoustical energy present in the environment.

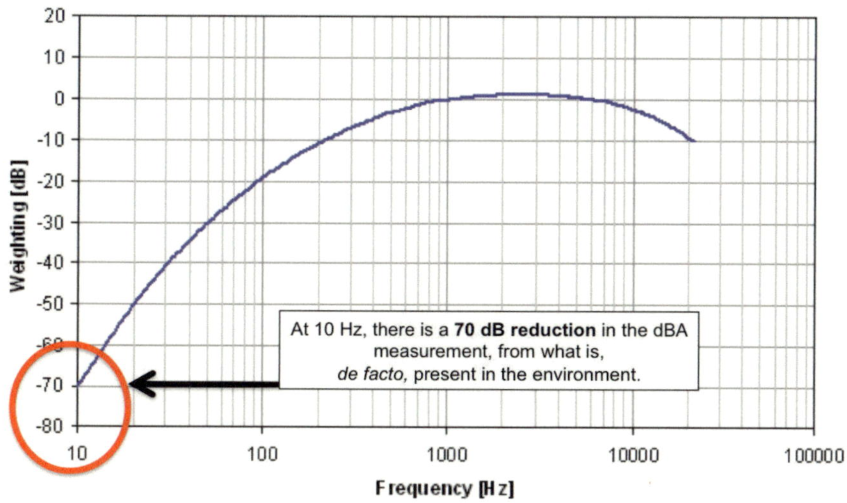

Figure 4.
Frequency response curve for the dBA metric applied to infrasonic frequency ranges, showing a 70 dB difference when evaluated at 10 Hz. Within these lower frequency ranges, the dBA metric will significantly underestimate the airborne acoustical energy present in the environment.

at these lower frequencies cannot be properly studied if the dBA metric is being used to characterize acoustical environments.

There is a shortage of studies that properly evaluate the biological response to infrasonic (≤20 Hz) or lower frequency (≤200 Hz) airborne pressure waves. Three important reasons for this have been provided above: the rudimentary segmentation of the entire acoustical spectrum into merely three "blocks" (compare to segmentation of the electromagnetic spectrum), the unsuitability of the dBA metric to quantify airborne acoustical pressure waves at these lower frequencies, and the ingrained notion that "what you can't hear can't hurt you." These major hindrances have been crystallized into mainstream science [3] and have served to significantly impede scientific inquiry and human health protection.

The goal of this chapter is to consolidate what is known on the biological response to airborne pressure waves occurring within the infrasonic and lower frequency ranges. A biomedical engineering approach is taken, whereby biological organisms are viewed as structures of composite materials, with significant viscoelastic components and organized in accordance with the principles of tensegrity architectures. When airborne pressure waves impact these types of structures, the biological response will depend on the type of biomaterial under study, it will exhibit anisotropic properties, and it will vary nonlinearly with exposure time. Depending on the physical properties of the airborne pressure waves (including time profiles) and on the biostructure under study, mechanical perturbations are relayed into cells and tissues through a variety of different pathways that, to date, still remain unclear.

2. Biomaterials and human anatomy

2.1 Viscoelasticity

Viscoelasticity is an attribute given to bodies that exhibit both viscous and elastic behaviors beyond the classical Hooke's elastic model [4]. Viscoelastic materials have three distinct properties not contemplated by Hookean models: creep, stress relaxation, and hysteresis. Most biological materials have viscoelastic behaviors.

In a Hookean (or purely elastic) material, total deformation depends on total load, and no further deformation occurs even if load is maintained. In viscoelastic materials, however, when sufficient stress is applied and maintained, they may continue to deform, even though stress load remains unaltered. This property is called *creep*.

In a purely elastic material, the strain within the material is constant throughout the application of the load; it does not vary with time, but only with the amount of applied stress. In viscoelastic materials, when stress is applied and maintained, strain can decrease with time. This property is called *stress relaxation*.

Consider repetitive or cyclical loads on materials. In purely elastic materials, periodic loads will not alter the stress-strain curve. The pathway taken by the material to deform is exactly the same pathway it takes to return to its original, equilibrium position. In viscoelastic materials, however, the return to equilibrium may be different than the pathway used to get to the point of deformation (The word pathway is here loosely used, and is meant to encompass all spatial, temporal and energetic components of these types of movements.) This property is called *hysteresis*.

2.2 Tensegrity structures

Many structures in the natural world are organized in accordance with the principles of tensegrity architecture—elements providing discontinuous compression are held together through elements of continuous tension [5]. **Figure 5** shows several examples of tensegrity structures.

Depending on the properties of the airborne pressure waves and biomaterial under study, the propagation of mechanical perturbations throughout these types of structures can reach long distances, without loss of structural integrity.

2.3 Cellular and tissue mechanotransduction

Cells and tissues are organized in accordance with the principles of tensegrity architecture [8, 9]. This means that in addition to biochemical signaling,

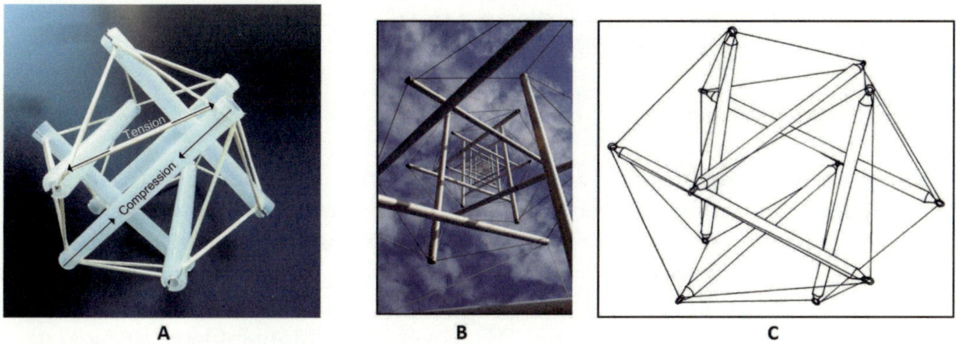

Figure 5.
Tensegrity structures. A. Model showing elements of continuous tension and discontinuous compression. B. Needle tower, by Kenneth Snelson, in the Hirshhorn sculpture garden (USA) [6]. C. Icosahedron, first designed by Buckminster Fuller in 1949 [7].

cells also communicate with their surroundings through mechanical signals. Mechanosensitive receptors exist on cell surfaces, and mechanosensitive junctions interconnect cells, thus forming tissues. Depending on the physical properties of the airborne pressure waves and biomaterials under study, external airborne mechanical perturbations can elicit a mechanical response, which, in a larger, macroscopic view, can lead to clinically pathological situations.

2.4 The fasciae

The fascia is a sheet of connective tissue that uninterruptedly extends from head to toe, suspended from the skeleton, and that provides the integrated supporting framework for maintaining anatomical and structural form [10, 11]. That external mechanical perturbations elicit responses at large distances away from the point of entry is a well-known concept among scientists and health professionals who study fasciae. When presented with external airborne pressure waves, fasciae can respond by changing their structural properties: *from a mechanical point of view, the fasciae are organized in chains to defend the body against restrictions. When a restriction goes beyond a specific threshold, the fasciae respond by modifying their viscoelasticity, changing the collagenic fibers, and transforming healthy fascial chains into lesioned chains* [10]. One of the fascia's key roles is that of shock absorption.

Connective tissue structures are ubiquitous forming all external surfaces of vessels, nerves, organs, and muscles, and at the cellular level, the extra-cellular matrix that surrounds and communicates with each individual cell. In addition to maintaining structural integrity, the fasciae are the first line of defense against external perturbations, playing important physiological roles in mobilizing the immune system.

3. Laboratorial studies, field studies, and biological outcomes

Studying the effects of infrasonic or lower-frequency airborne pressure waves on biological structures is a very complex undertaking, whether it be on cell cultures, on animal models, or on human populations. Laboratorial studies, occupational field studies, and residential field studies all have their own strengths and weaknesses. When the latter go unrecognized, however, experimental design flaws can ensue. In this section, the attributes of these different experimental setups are discussed, and their weaknesses and strengths are explored. Together with the preceding section, this serves as a preamble to Section 4, where the results of experimental studies are described in detail.

3.1 Laboratorial studies

Laboratories where infrasonic and lower frequency airborne pressure waves can be applied in a controlled manner are in short supply worldwide, and those that do exist are mostly associated with military installations. Laboratories emitting airborne pressure waves with infrasonic and lower frequency components cannot be randomly placed within residential environments; issues with neighbor disturbance and public health would curtail its use. Moreover, the equipment used to generate the airborne pressure waves is, typically, very large and very expensive, and few sectors of society (other than military or space exploration industries) would have the need for an extensive use of these types of installations.

In these laboratory settings, continuous or pulsed-trains of single-tone airborne pressure waves can be applied, as well as, broadband exposures that can be accurately characterized. The fact that exposure times and acoustic parameters can be precisely controlled is one of the strengths of laboratorial studies, allowing for continuous time exposures, or occupationally simulated exposure schedules. Immediate (hours or days) versus long-term (weeks or months) effects can also be explored.

There are numerous types of biological outcomes that can be studied under laboratorial conditions. Light-, electron- and atomic-force microscopy can be used to study cellular and tissue structural properties, as well as their chemical composition and content of bio-reactive elements. Polymerase chain reaction (PCR) techniques can provide information on messenger RNA (mRNA) expression, allowing for the identification of key pathways. With pharmacological intervention or gene knockout specimens, specific signaling molecules and pathways involved in the elicited responses can be pinpointed. Additionally, control populations for comparison are fairly easy to achieve—they are simply not subjected to the laboratorial exposures.

3.2 Occupational field laboratories

Occupational environments are exceptional field laboratories, as both short-term (several months) and long-term (years) effects can be investigated in more realistic acoustic environments. Typically, different workstations have different acoustical features that can greatly depend on different machinery regimens. For occupational field laboratories, acoustical characterizations of the workplace(s) must be comprehensively undertaken and time exposures to each type of environment should be scored.

Exposure times at work must be differentiated from exposure times away from work, i.e., when the work shift ends, workers leave the field laboratory, but additional exposures to infrasonic or lower frequency airborne pressure waves may be incurred (e.g., recreational, transportation). These must be documented. Significant confounding factors may be introduced unless each subject's residential area is scrutinized and prior-exposure histories probed for fetal, childhood, and adolescent exposures.

Possible biological outcomes within occupational field studies are more limited when compared to laboratory exposures. Noninvasive testing can be imprecise, and the minimally invasive testing (such as a blood chemistry analysis, X-ray, or MRI) may also not be sufficiently precise to yield relevant data. It is also the case that scientific knowledge on relevant biological outcomes that can be noninvasively evaluated in exposed humans is still absent or, at best, very incomplete.

Survivorship bias is a well-known confounding factor in human population studies. In occupational environments, workers with more time on-the-job are those who have survived throughout the years of professional activity, while workers with less time in professional activity may exhibit more severe biological outcomes. This phenomenon is often misinterpreted leading to inconclusive or erroneous conclusions.

Control populations for exposures to infrasonic and lower frequency airborne pressure waves have been a very difficult proposition, given the ubiquitous nature of this stressor. One of the solutions to this profound problem is the scoring of subjects into different groups as per their exposure. Within this context, control groups are composed of individuals who have the least amount of cumulative (prior and present) exposure, and not of individuals with zero exposure.

Different professions can provide different field laboratories, both in terms of acoustic environment and time exposure schedules. For example, long-haul truck drivers are typically exposed for more than 8 hours daily and, oftentimes, sleep in the truck while it is idling, or while refrigeration systems are continuously operating. Workers onboard ships, submarines, offshore oilrigs, aircraft, and spacecraft (for example) can be exposed to significant amounts of infrasonic and lower frequency airborne pressure waves for weeks or months at a time. The wealth of information waiting to be gleaned from these types of field laboratories is breathtaking.

3.3 Residential field laboratories

Field laboratories in urban, suburban, and rural residential settings are generally designed to investigate environmental health effects due to human-made infrasonic and lower frequency airborne pressure waves. Typically, these sources are associated with industrial complexes or infrastructure that, in turn, are usually linked with important economic interests. In general, the amount and type of infrasonic and lower frequency airborne pressure waves contaminating a home will depend on the machine operation and/or the use of the infrastructure. For example, in most urban and suburban areas, airports must close down between the hours of midnight and 5 am. Some factories do not have night shifts and therefore also have daily shutdown periods. Large refrigeration units, hydroelectric dams, and large volume highways, however, must be kept running 24/7 and can also be viewed as continuous sources of infrasonic and lower frequency airborne pressure waves. Wind turbines are the latest addition to these type of sources although they are almost exclusively within rural areas.

Comprehensive characterization of the acoustic environments in the different residential areas must be undertaken (e.g., master bedroom, children's bedrooms, living-lounge areas), since room-resonance phenomena can significantly modify the acoustic environment that is originally being induced and driven by external, incoming airborne pressure waves. Additionally, wind can also influence the spectrum, intensity and type of infrasonic and lower frequency airborne pressure waves that exist within a room. This differentiation is readily achieved with proper acoustic evaluations.

Residential exposure times are much more difficult to control, as they can differ from room to room and on an hourly basis. Moreover, subjects may also be sleeping within the "contaminated" environments, which can severely aggravate biological outcomes. If exposure is concomitantly occurring during sleep and waking hours (e.g., homemakers, workers from home, farmers), then biological outcomes may be further aggravated. Leaving the home can be equated with a biological recovery period (i.e., nonexposure period).

Short-, medium- and long-term effects can be studied in residential settings when the implementation of a new infrastructure or industrial complex is known to be coming to the area. Biological outcomes should strive to be either noninvasive or minimally invasive, and prior-exposure histories are fundamental for achieving useful statistical data.

4. Past relevant studies

Numerous studies conducted over the decades have shed light on the biological response to infrasonic and lower frequency airborne pressure waves and associated symptomatic complaints. Due to space limitations, this discussion will only deal with some of the vascular and collagenous abnormalities, cardiomyocyte changes, and the hippocampus responses, as induced by different types of exposures. For reasons explained in the section "Introduction," all studies using the dBA metric have been eliminated from consideration (with one exception in an occupational setting). Selected studies mostly focus on the cellular and tissue changes observed in laboratory, occupational, and residential settings, using light and electron microscopy. The sequence in which the studies are presented does not follow the classical anatomical order.

4.1 Vascular changes

In the mid-1960s, within a military setting, the immediate exposure to 10–60 Hz, at 118–140 dB, for 2 minutes, induced disturbances of the visual field as reported by all five human subjects [12]. In 1985, laboratorial animal studies exposed rats to tonal 8 Hz at 100–140 dB, 3 hours daily, for 5, 10, 15, or 25 days, and examined the blood and lymph networks of the palpebral (eyelid) and bulbar (eye globe) conjunctiva. *Day 5*: narrowing of all parts of the conjunctiva blood network was observed, with decreased blood capillary lumens. Capillaries, precapillaries, and arterioles were twisted, and blood component agglomerations were identified in venous vessels. *Day 10*: conjunctiva capillaries were twisted and large vessel diameters were decreased. *Day 15:* blood and lymph vessel tonus had changed, and stagnation was present. *Day 25:* failure of tissue homeostasis was aggravated. Capillary penetration was increased, as seen through tissue enlargement, and significant agglutination was observed in the large vessels [13].

In a similar study, animals were exposed to 8 Hz at 100 dB, or to 16 Hz at 100 dB, 3 hours daily, for 1 month. Clinical and morphological evaluations were conducted at days 3, 7, 15, 30, and also post-exposure at days 30, 60, and 90. *Day 3:* clinical changes were not observed, but morphological changes were present: edema in the upper and middle areas of the eyelid derma and heterogeneous blood filling of vessels with extra-vascular erythrocytes were also observed. Fine focal hemorrhages were identified under the corneous layer of the eyelid. Sclera exhibited edema, and blood vessels were filled heterogeneously with stasis and extra-vascular intraconjunctive hemorrhages. In the 8-Hz group, moderate edema was present near the optical nerve, and the 16-Hz group exhibited perineural hemorrhages in the optical nerve. *Day 7:* in both groups, conjunctiva blood vessels had expanded and arteries in the oculus fundus were narrower and twisted. Eyelid edema of the derma was identified in both groups. The most pronounced vascular changes were found in the eyelid conjunctiva: stasis, edema, and pericapillary hemorrhages. Sclera capillaries were overfilled with blood and extra-vascular hemorrhages were observed. *Day 15:* in both groups, conjunctiva vessels were narrower and twisted, and ocular globe conjunctiva exhibited nonvascularized areas. Vascular changes as seen previously were more expressed: edema, paresis state in capillaries (erythrocyte stasis), and extra-vascular erythrocytes. The iris exhibited narrower vessels. *Day 30:* narrowed and twisted vessels were clinically detected, with ocular fundus arteries and veins significantly narrowed and twisted, more pronounced in the 16-Hz group. In the eyelid conjunctiva, derma exhibited the same vascular changes seen before: edema and erythrocyte stasis. Sclera arteries and veins were larger, overfilled with blood, and with the presence of extra-vascular focal and diffuse hemorrhages with conjunctiva involvement. At all time points, the 16-Hz group disclosed more destruction than the 8-Hz group. *Day 60 (30 days post-exposure)*: clinical evaluations revealed less twisted and narrow arteries and veins, but morphological recovery was slower. In the 8-Hz group, moderate regeneration was observed in the eyelid conjunctiva epithelium. In the 16-Hz group, predominant retinal damage persisted. *Day 90:* no clinical changes were observed in either group [14].

Within an occupational setting (reinforced concrete factory), vessel changes in the palpebral and bulbar conjunctiva, and in the retina, were investigated among 214 workers (age range: 20–58 years), with 1–30 years of employment. Workers were divided into two groups:

- *Control group* (n = 54): not occupationally exposed to significant levels of infrasonic and lower frequency airborne pressure waves.

- *Exposed group* (n = 160): tonal 8 and 16 Hz at 96–100 dB, simultaneously with non-tonal 20–500 Hz at 91–93 dBA.

The exposed group was divided into subgroups as per years of professional activity. **Table 1** describes each subgroup and the vessel abnormalities found. No such abnormalities were found in the control population [14].

Within a different occupational setting (aircraft industry), ocular changes were studied in 23 male workers (average age: 42, range: 32–58 years). Lesions

Occupational exposure time	**1–2 yrs**	**3–10 yrs**	**11–20 yrs**	**20–30 yrs**
Number of workers	21	84	36	19
Palpebral and bulbar arteries (%)				
Enlarged	0	82	8	0
Narrow	0	17	91	100
Twisted	0	80	100	100
Retinal arteries (%)				
Enlarged	0	0	0	0
Narrow	0	91	100	100
Twisted	0	90	100	100
Retinal veins (%)				
Enlarged	0	87	11	0
Narrow	0	13	88	100
Twisted	0	75	97	100

Table 1.
Percentage of abnormal vessel changes seen in the palpebral and bulbar conjunctiva and retina among occupationally exposed workers [14].

were observed in the blood-retinal barrier in 19 workers (lesion types: 13 inactive, 2 active, 4 mixed). Choroidal circulation was altered in 14 workers (late perfusion with chronic features). Changes in retinal circulation were observed in four workers (type: 1 occlusive, 1 exudative, 2 mixed). Three workers presented with optic neuropathy (1 papillitis, 2 optic atrophy), and one exhibited sensorial retinal macular detachment [15]. The immediate effects of tonal exposures with 8 Hz at 130 dB, 2 hours daily, for 1, 7, 14 and 21 days, also revealed a breakdown of the blood-retinal barrier in the rat eye [16].

These studies strongly suggest that under the impact of infrasonic and lower frequency airborne pressure waves, a vascular response is mounted by ocular structures and could be related to decreased visual acuity in workers. Data in **Table 1** seem to indicate that, as exposure time progressed, vessels that were initially enlarged ceased to exist, apparently being replaced with narrower and twisted vessels. Enlarged vessels usually suggest the need for an increased blood supply. However, given the sustained mechanical insult, making the vessels narrower and twisting them throughout the structures may, in fact, reflect a more efficient blood delivery system.

This concept is further reinforced by the observation of narrow and twisted blood vessels in the gastric mucosa of rats, exposed to non-tonal, occupationally simulated (aircraft industry) acoustic environments characterized as 6.3–25 Hz at 70–90 dB and 40–500 Hz at 90–100 dB. Continuous exposure was applied, and evaluations occurred at 1, 3, 5, 9, and 13 weeks. In 3–5 *weeks*, the gastric submucosal layer exhibited significantly increased thickness, when compared to non-exposed controls. This increased thickness was due to the proliferation of type IV collagen. Arterial walls disclosed significant intima and media thickening, ruptured internal elastic lamina, and thrombotic changes. In *9–13 weeks,* neoangiogenesis was observed, with the appearance of tortuous and twisted vessels. The authors concluded that, in the stomach, continuous exposure induced fibrosis that could be linked with neoangiogenesis, since collagen type IV is also an early marker of neoangiogenesis [17]. One of the earliest studies investigating the long-term effects of airborne pressure waves on gastric complaints was conducted in 1968, in a residential setting where changes in gastric function were associated with aircraft noise [18]. Within occupational settings, an increase in gastric complaints was documented among boiler-plant workers, 2 years after the implementation of mandatory hearing protection devices [19]. Among aircraft industry workers, gastrointestinal problems were among the earliest to appear after 1–4 years of professional activity [20].

Vascular changes were also identified in the liver structures of animals exposed to 2, 4, 8, or 16 Hz, at 90–140 dB, 3 hours daily, for 5–40 days. Exposures to 2 or 4 Hz induced less damage than exposures to 8 and 16 Hz. *Single, 3-hour exposures:* with 2 or 4 Hz and 90 dB, no changes were observed in the hepatic structures, while at 100–110 dB, liver parenchyma disclosed single fine hemorrhages. At 120 dB, increased arterial wall diameters were observed, as well as capillary lumen expansion, indicating the development of ischemia. At 130–140 dB, the number of hemorrhagic events increased, as did the number of affected hepatocytes. With 8 or 16 Hz exposures, damaged hepatocytes were present in the ischemic and non-ischemic areas. *Days 5–15:* more pronounced hepatocyte changes were seen. *Days 25–40*: a gradual death of changed hepatocytes was observed [21].

Hemorrhagic events in the lung were documented as early as 1969, within the Soviet and US space exploration studies, in dogs exposed to occupationally simulated (spaceflight) wide-band frequency range at 105–155 dB, for 1.5 or 2 hours. Hemorrhages up to 3 mm in diameter were observed beneath the pleura. As exposure time and decibel level increased, the number of hemorrhages increased but never exceeded 3 mm in diameter. Microscopic analyses of the hemorrhagic sections disclosed ruptured capillaries and larger blood vessels [22]. In a laboratory setting, rats received tonal exposures to 2, 4, 8, or 16 Hz at 90–140 dB, 3 hours daily, for 40 days. Analysis time points were conducted after 3 hours, at 5, 10, 15, 24, and 40 days of exposure, as well as during post-exposure times. *Single, 3-hour exposures:* with 2 or 4 Hz at 90–110 dB, mosaic hemorrhages were observed under the pleura, covering the entire lung surface. With 8 Hz at 110 dB, more hemorrhagic expression was observed. With 8 or 16 Hz at 120–140 dB, larger hemorrhagic foci were disclosed. Within the alveolar capillary network and postcapillary venules, vessel diameters were increased with 2 or 4 Hz at 90–110 dB, leading to large hemorrhages and perivascular edema. Erythrocyte overflow in alveolar capillaries was observed with 8 or 16 Hz at 110 dB. With 8 or 16 Hz at 120 or 140 dB, lung tissue exhibited large hemorrhagic foci in the connective tissue septa of the bronchi-pulmonary segments. In all exposure types, capillary changes were followed by alveolar epithelium desquamation and basal membrane denudation. *Longer exposures:* with 8 Hz at 120 dB, acinuses became filled with erythrocytes, and interstitial hemorrhagic foci caused a strong deformation of the respiratory bronchioles. With 8 or 16 Hz at 140 dB, ruptured vascular walls were observed leading to decreased alveolar lumen [23].

The highly invasive bronchoscopic evaluation with biopsy was performed among a group of volunteer subjects, with occupational or residential exposures to infrasonic and lower frequency airborne pressure waves, as detailed in **Table 2**.

Bronchoscopic observations in all patients revealed small submucosal, vascular-like lesions ("pink" lesions), located distally in both tracheal and bronchial trees, and uniformly distributed bilaterally near the spurs. Biopsies were performed on the abnormal mucosa (pink lesions) and on the apparently normal mucosa (outside of the pink lesions). In the non-pink areas, some vessel wall thickening was visible. In the pink areas, the basal membrane disclosed abnormal neovascularization, with thickened blood vessel walls and scarce lumen. No gender differences were identified [24].

4.2 Collagen and connective tissue

Collagen, composed of triple-helix tropocollagen chains, is the most abundant protein in the human body, a key component of the fasciae, and is produced by fibroblast cells. It has long since been considered as the "steel" of the human body [25], but its energy storage capacity has been shown to be 10 orders higher than in spring steel [26]. Different types of collagen have different mechanical properties. Type IV collagen (increased in the exposed gastric mucosa [17]—see above), is organized into X-shaped structures and is commonly found in the basal membrane of arterial walls, hence its increased expression during angiogenesis.

In *day 5* of the eyelid-and-bulbar-conjunctiva animal studies (see above [13]), collagen fibers in the connective tissue were enlarged, as were some fibroblast nuclei; on *day 10*, adipose cells in the connective tissue had been redistributed and positioned in the vascular areas of the conjunctiva. In the second animal study described above [14], *day 3* included edema of the sclera causing separation of collagen filaments in the 16 Hz group, and by *day 7,* this was observed in the 8-Hz group as well; *day 15:* focal and disseminated disorganization of sclera collagen fibers was observed in both groups; *day 30:* homogenization and disorganization of collagen in the derma while, in the sclera, collagen fibers were persistently separated due to edema, with some undergoing dystrophic and necrotic changes. Slow regeneration was observed during the post-exposure periods.

In the lungs of dogs studied within the scope of space exploration (see above [22]), focal enlargement of the alveoli involved the stretching of connective tissue

Profession/type of exposure	Gender	Age	Smoking
Aircraft technician	Male	48	Mild
Aircraft technician	Male	52	No
Aircraft technician	Male	59	Mild
Combat pilot	Male	61	No
Helicopter pilot	Male	59	Moderate
Aircraft pilot	Male	54	No
Merchant marine	Male	37	No
Military helicopter nurse	Female	56	No
Flight attendant	Female	36	No
Flight attendant	Female	39	No
Flight attendant	Female	40	No
Homemaker	Female	54	Mild
Homemaker	Female	59	No

Table 2.
Description of subjects who received bronchoscopic evaluations with biopsy [24].

structures of alveoli walls. In the biopsy images of the bronchoscopic study (see above [24]), non-pink areas disclosed a thickened basement membrane with abnormal amounts of collagen, while the pink areas disclosed an even thicker membrane with very large amounts of collagen. The abnormal neovascularization was embedded within collagen bundles. Retraction of structures neighboring the collagen fibers was not observed. A marked reinforcement of the cytoskeleton and intercellular junctions was seen in the pink areas, as compared to non-pink areas. The five individuals that disclosed images of collagen fiber degeneration and disruption also tested positive for antinuclear antibodies.

Under an occupationally simulated acoustic environment, characterized as 20–200 Hz at 70–90 dB (aircraft industry), and occupationally simulated exposure schedules (8 hours daily, 5 days weekly, weekends in silence), focal interstitial fibrosis was found in the lung parenchyma of rats after a cumulative 4000-hour exposure. Additionally, thickened alveoli walls and dilated alveoli were observed [27]. Tracheal epithelium in similarly exposed rats disclosed significant subepithelial fibrosis [28, 29], and with longer occupationally simulated exposures, the subepithelial layer became composed of hyperplastic collagen bundles, some with a degenerative pattern. Cellular edema was also observed [28, 30].

Within an occupational setting (aircraft industry) and investigating long-term outcomes, high-resolution CT scans of the lungs and respiratory function tests were provided to 21 nonsmoker male workers, who were divided into two groups: with (n = 7, average age: 42) and without (n = 15, average age: 36) complaints of airflow limitations. There was a significant relationship between the presence of symptoms and images of lung fibrosis through the CT scan. No differences existed among the groups when comparing the percentage of predicted values of lung function [31].

Fasciae abnormalities have been most prominently studied in the pericardia of exposed workers, subsequent to autopsy findings in an aircraft industry worker that disclosed a grossly thickened pericardium [32]. Pericardial morphological changes were studied among 12 male workers: three aircraft technicians, four fixed-wing aircraft pilots, four helicopter pilots, and one long-haul truck driver. Pericardial samples were removed with informed consent of the patient and Ethics Committee approval, at the beginning of cardiac surgery (prescribed for other reasons by the National Healthcare Service). In all cases, there were no visual adherences, or inflammatory aspects and pericardia were grossly thickened. The classical, three pericardial layers were identified: serosa, fibrosa, and epipericardium. However, in all cases, the fibrosa had split in two and, in between, a new layer of loose tissue was observed, consisting of vessels, nerves, arteries, and lymphatics surrounded by adipose tissue. Both fibrosa layers were composed almost entirely by wavy, interwoven collagen bundles, surrounded by numerous cytoplasmic extensions (whose mother cell was difficult to identify), and interspersed with some elastic fibers. The new, loose tissue layer sandwiched in between the split fibrosa contained blood and lymphatic vessels, adipose tissue, and nerves. Both the loose tissue layer and the fibrosa layers contained macrophages and vascular hyperplasia, also seen in lymphatic vessels [33–36]. Pericardial and cardiac valve thickening has also been confirmed through echocardiography studies in occupational settings (aircraft [37] and commercial-airline industries [38]), with thickness increasing with increasing exposure time. In residential settings, pericardial and valve thickening [39] and increased arterial stiffness [40] were observed in populations chronically exposed to military-training exercises [39], and transportation systems [40].

4.3 Heart cells and tissues

In 1983, electron microscopy techniques were used to study animal myocardia exposed to single and multiple infrasonic exposures of 4–16 Hz at 90–150 dB, 3 hours daily, for 45 days, and post-exposure time points were included. No changes were observed with single exposures at 4–6 Hz and at less than 100 dB, when compared to non-exposed controls. *Single exposure with 4–10 Hz at 120–125 dB:* induced decreased arterial diameter and capillary expansion, with resulting focal ischemia. Images of intracellular myocytolysis were frequently found. These processes were reversible. *Multiple exposures with 4–10 Hz at 120–125 dB for 5–25 days*: ventricle fibrillation and subsegmental contractures in ischemic foci were identified. Myofibril fragmentation was observed in the Z-line, sarcoplasmic reticulum structures were absent, cell nuclei were deformed, and chromatin was found accumulated under the nuclear membrane. *post-exposure*: intracellular regeneration was concomitant with damaged cells. In surviving cells, mitochondria were increased in number and size, and both myofilaments and sarcoplasmic reticulum elements were being created. Intracellular regeneration was slow and ended with the creation of Z-lines, after which myofibrils became normal and myocardiocytes completely recovered. *Single exposure with 10–15 Hz at 135–145 dB:* more pronounced myocardial damage, with partial death of myocardiocytes, resulting in myocardiocyte dystrophy. Damaged cells included chromatin condensation and redistribution to the nuclei membrane. Less damaged cells regenerated after 5–10 days post-exposure. *Multiple exposures with 10–15 Hz at 135–145 dB:* persistent myocardial ischemia related to vascular changes and accompanied by cardiocyte damage. After 15–25 days post-exposure, recovered cells began functioning normally despite the presence of abnormal structures within the cellular cytoplasm, namely, giant mitochondria [41].

Cardiac injury was studied in rat cardiomyocytes exposed to tonal 5 Hz at 130 dB, 2 hours daily, for 1, 7, or 14 days. *Days 1–7:* SERCA2 (sarcoplasmic reticulum Ca^{2+} ATPase 2, an enzyme with calcium-transporting properties and involved in the decomposition of ATP into ADP) was significantly increased, and swollen mitochondria were observed in the cardiomyocytes. *Day 7:* SERCA2 was significantly decreased and an increased number of swollen mitochondria were observed. *Day 14:* SERCA2 was significantly decreased and platelet aggregation was found in the intercellular substance. Intercellular calcium ion (Ca^{2+}) concentration significantly increased with increasing exposure time [42]. With similar exposure protocols, another study repeated the SERCA2 and intercellular Ca^{2+} concentrations, but also included evaluations of the expression of whole cell L-type Ca^{2+} currents (WLCC) and the mRNA expression of a subunit of the L-type Ca^{2+} channel (LCC). SERCA2 and intercellular Ca^{2+} concentrations behaved as described immediately above, while the expression of WLCC and mRNA expression of LCC increased with increasing exposure time [43].

For three continuous months, rats were exposed to non-tonal, occupationally simulated (aircraft industry) acoustical environments characterized as 6.3–25 Hz at 70–90 dB and 40–500 Hz at 90–100 dB. Ventricular cardiac muscle and interstitial fibrosis were quantified and compared to non-exposed controls. Exposed rats disclosed a 97.5% increase in fibrosis in the left ventricle, an 81.5% increase in the interventricular septum, and an 83.7% increase in the right ventricle. No significant differences were found in the mean values of cardiac muscle in the left and right ventricles, when compared to non-exposed controls. However, the fibrosis-to-muscle ratio was significantly higher in the exposed rats, indicating significant ventricular myocardial fibrosis [44].

In another study, rats were exposed to a non-tonal, occupationally simulated (textile mill) environment rich in infrasonic and lower frequency components, under an occupationally simulated schedule (8 hours daily, 5 days weekly, weekends in silence), for 1, 3, 5, and 7 months. Ventricular coronary artery caliber, artery wall thickness, and size of arterial perivascular tissue were quantified in a total of 130 arteries (61 exposed and 69 controls). No changes were observed in arterial lumen caliber, and in arterial wall thickness, when compared to non-exposed controls. Perivascular tissue was more prominent in the exposed samples and seemed to exhibit fibrotic development. Lumen-to-wall ratio showed no differences, while wall-to-perivascular-tissue ratio showed a significant increase, as compared to non-exposed controls [45].

In animals exposed to 2–20 Hz peaking at 114 dB, for 28 continuous days, ventricular arteries were studied as to the dimensions of lumen, wall, and perivascular space. An additional group of animals received the same exposure but were treated with dexamethasone (a corticosteroid). Blind evaluation of 31 arteries disclosed increased perivascular spaces in the exposed groups, reflected in the significantly reduced wall-to-perivascular-space ratio, as compared to non-exposed controls. No changes were observed in the lumen-to-wall ratio. With dexamethasone treatment and exposure, no differences were observed in the wall-to-perivascular-space ratio, as compared to controls, suggesting an underlying inflammatory mechanism [46].

Gap junctions are a fundamental component of intercellular communication, allowing inorganic ions and small water-soluble molecules to pass directly from one cell's cytoplasm to another. Gap junctions are formed by protein complexes (connexons) each composed of six subunits made of the protein connexin. Cardiac connexin43 (Cx43) is a component of gap junctions, and its reduction in combination with increased collagen deposition and interstitial fibrosis has been associated with ventricular arrhythmias [47]. Within this context, rats were exposed to non-tonal, occupationally simulated (aircraft industry) acoustical environments characterized by 6.3–25 Hz at 70–90 dB and 40–500 Hz at 90–100 dB, for three continuous months. Immunohistochemical quantification of Cx43 was conducted on the left ventricle, interventricular septum, and right ventricle. Significantly decreased Cx43-to-muscle ratios were found in the exposed rats, as compared to non-exposed controls, suggesting the possibility of arrhythmogenic consequences [48].

4.4 The hippocampus

Prior studies have shown that the hippocampus is involved in learning and memory impairment, such as that seen in rodents after infrasound exposure [49]. The hippocampus—located between the cerebral hemispheres and the brainstem—was classically considered as part of the limbic system. The hippocampus proper is divided into four regions (CA1, CA2, CA3, and CA4), each with different input and output pathways. The Dentate Gyrus (DG) is an additional hippocampus structure and that contributes to the formation of new episodic memories, and spontaneous exploration of novel environments. In the central nervous system (CNS), neuroglia consists of the non-neuronal cells (oligodendrocytes, astrocytes, ependymal cells, and microglia) and is often referred to as the connective tissue of the brain. Glial cells surround neurons to hold them in place, supply them with oxygen and nutrients, insulate them from one another, destroy pathogens, and remove dead neurons.

Glial fibrillary acidic protein (GFAP) is an intermediate filament protein expressed by numerous cells within the CNS, and although its exact function remains unknown, it appears to be involved in maintaining the mechanical strength of astrocytes. The expression of GFAP was studied in the brains of mice exposed to 16 Hz at 130 dB, 2 hours daily, for 1, 7, 14 21, or 28 days. GFAP expression was increased in the hippocampus, cortex, and hypothalamus in a time-dependent manner [50].

Corticotrophin releasing hormone (CHR) is a peptide hormone involved in the stimulation of the pituitary synthesis of ACTH (adrenocorticotropic hormone) as part of the hypothalamic-pituitary-adrenal axis' response to stress. Corticotrophin releasing hormone-receptor 1 (CHR-R1) has wide expression in the CNS. It plays important roles in fear learning and consolidation in the amygdala, in stress-related modulation of memory function in the hippocampus, and in arousal regulation in the brainstem. Prior studies showed that infrasound exposures caused an upregulation of CRH and CRH-R1 in neurons of the hypothalamic paraventricular nucleus [51]. Recent studies have also shown that CRH is expressed in activated microglial cells [52]. Within this context, rats and *in vitro* cultured microglial cells were exposed to 16 Hz at 130 dB for 2 hours, after which changes in CHR-R1 were examined. *In vivo* exposure disclosed activation of microglial cells and an upregulation in the expression of CRH-R1 in the hypothalamic periventricular nucleus. *In vitro* exposure disclosed that, in the absence of neurons, microglial cells were activated and CRH-R1 expression was upregulated. These data suggest that both neurons in the hypothalamic periventricular nucleus and microglial cells are effector cells for infrasound-elicited responses [51].

The transient receptor potential cation channel, subfamily V, member 4 (TRPV4) protein acts as a calcium channel that is also mechanosensitive. It plays important roles in the systemic regulation of osmotic pressure by the brain, in skeletal growth and structural integrity, in airway and lung function, retinal and inner ear function, and in pain. Animals were exposed to 8 or 16 Hz at 90, 100 or 130 dB, 2 hours daily, for 14 days. Rat learning and memory abilities were most severely impaired with 16 Hz at 130 dB at days 7 and 14, with prominent loss of hippocampal CA1 neurons, as compared to non-exposed controls. Significant astrocyte and microglial activation was seen in the hippocampus after days 1 and 7, and before neuronal apoptosis became evident. *In vivo* pharmacological intervention causing the inhibition of glial activation protected against neuronal apoptosis. *In vitro*, exposed glial cells released proinflammatory cytokines, a key factor for neuronal apoptosis. In both *in vivo* and *in vitro*, expression levels of

TRPV4 were increased as compared to non-exposed controls. Pharmacological or knock-out intervention of TRPV4 in cultured glial cells decreased the levels of inflammatory cytokines and attenuated neuronal apoptosis. This study also demonstrated the involvement of calmodulin and protein kinase C signaling pathways in the response to infrasonic exposures. These data suggest that TRPV4 expressed by glial cells is potentially a key factor in infrasound-induced neuronal impairment [53].

Neonatal rat hippocampal astrocyte cultures were exposed to 16 Hz at 130 dB for 15, 30, 60, 90, 120, and 240 minutes. Extra-cellular glutamate levels increased with increasing exposure time, and at 90 min, there was a 100% increase over baseline. The astroglial expression of Cx43 (connexin43—see above) was increased, as compared to non-exposed controls, as was the synthesis of Cx43 mRNA. Through additional evaluations using pharmacological and knock-out interventions, the authors concluded that infrasonic exposures induced astrocytes to release glutamate, and that Cx43 gap junctions were required for the exposure-induced glutamate release [54].

The endocannabinoid system includes lipid-based retrograde neurotransmitters, expressed throughout the CNS, and involved in fertility, pregnancy, pre-and postnatal development, appetite, pain-sensation, mood, and memory. Animals were exposed to 16 Hz at 130 dB, 2 hours daily, for 14 days. Cannabinoid (CB) receptors 1 and 2 in the CA1 hippocampal region of the exposed rats were downregulated in a time-dependent manner, as compared to non-exposed controls. Apoptotic cells in the CA1 only became obvious after day 5, and cell death coincided with the decreased expression of CB receptors. Through pharmacological intervention, activation of CB receptors significantly reduced the number of apoptotic cells, ameliorated the behavior performance of exposed rats, and reduced the infrasound-elevated levels of proinflammatory cytokines. These data suggest that CB receptors could potentially serve as promising targets for future treatments against infrasound-induced injury [55].

Fibroblasts synthesize extracellular matrix (glycosaminoglycans, reticular, and elastic fibers) and collagen, and, in addition to their structural role, fibroblasts are also important for mounting the immune response to tissue damage. Fibroblast growth factors (FGF) signal through fibroblast growth factor receptors (FGFR). The fibroblast growth factor 2/fibroblast growth factor receptor 1 (FGF2/FGFR1) signaling pathway was investigated in animals and in cultured astrocytes, exposed to 16 Hz at 150 dB, 2 hours daily, for 1, 3, or 7 days. In both experimental models, astrocyte activation increased with exposure time and astrocyte-expressed FGFR1 was downregulated as compared to non-exposed controls. Pharmacological intervention using FGF2 exerted an inhibitory effect on infrasound-induced astrocyte activation, inhibited the elevation of proinflammatory cytokines, upregulated the expression of FGFR1, and alleviated neuron loss in CA1 hippocampus region. Inhibition of the FGF2/FGFR1 pathway aggravated astrocyte-mediated inflammation after infrasonic exposure. The authors concluded that astrocyte-mediated inflammation was involved in infrasound-induced neuronal damage and that the FGF2/FGFR1 pathway played a key role [56].

In a laboratory setting, rats were exposed to tonal 8 Hz at 140 dB, 2 hours daily, for 3 days. A post-exposure, 1-week time point was also established. Significant damage of hippocampus morphology was observed in exposed rats, and recovery was seen after 1 week of post-exposure. Neuronal apoptosis was significantly increased after 24- and 48-hour exposures, as compared to non-exposed controls, and then decreased after 1 week post-exposure. Expression of heat shock protein 70 (HSP70) peaked at 24 hours and was decreased at 48 hours [57].

5. Conclusions

Exposure to infrasonic and lower frequency airborne pressure waves can cause cellular and tissue damage depending on frequency, dB-level, and exposure time, while the viscoelastic properties inherent to biological tissues impart a nonlinear response to this type of acoustic stressor. The complex mechanosensitive and biochemical cellular signaling pathways mediating this cellular damage have not yet been pinpointed, although fasciae structures and connective tissues (including the neuroglia) seem to be the most sensitive under longer term exposures. Immediate exposures appear to induce inflammatory processes that do not seem to be maintained with longer exposures.

Widespread vascular involvement (not limited to the biological structures addressed herein) was observed in palpebral and bulbar conjunctiva and retina, gastric mucosa, liver structures, lungs, pleura and tracheae, alveoli, pericardia, and coronary arteries. This vascular response may (unsuspectingly) be the underlying cause of many symptomatic complaints. Cognitive deficits oftentimes documented within residential field laboratories may not merely be due to sleep deprivation, but also to hippocampal neuronal damage. Fasciae morphogenesis speaks to the demand on the whole-body structural integrity elicited by this type of external mechanical insult, while collagenous growths and hemorrhagic events of a focal nature may reflect concomitant resonance phenomena.

Recovery periods are not linear, and 2-hour daily exposures imply a 22-hour nonexposure period. This presents a problem for continuous exposures, such as those encountered in some professional activities and most residential environments. The underlying objectives of most of the studies discussed herein are related to occupational exposures and do not consider continuous exposures at less than 90 dB, nor are pressure pulsed trains presented within the laboratorial acoustic environments. In residential environments, however, these attributes are often present. The simulation of residential exposures does not appear to have yet been integrated into laboratory settings and protocols.

The whole-body response also elicits the immune system, affects organs of the reproductive system, changes receptor cells in the vestibular semicanals and auditory cochlea, and induces genotoxic effects, including teratogenesis. This is a pioneering field of science, still in its infancy and urgently requiring scientists from multidisciplinary areas of study because, ultimately, the health of human populations and their offspring must be protected.

Conflict of interest

None.

Author details

Mariana Alves-Pereira[1]*, Bruce Rapley[2], Huub Bakker[3] and Rachel Summers[3]

1 Lusófona University, Lisbon, Portugal

2 Atkinson and Rapley Consulting, Palmerston North, New Zealand

3 Massey University, Palmerston North, New Zealand

*Address all correspondence to: m.alvespereira@gmail.com

References

[1] WHO. The Burden of Disease from Environmental Noise. Copenhagen: WHO Europe; 2011

[2] Dirac Dirac Delta Science & Engineering Encyclopedia. A-weighting. 2017. Availble from: diracdelta.co.uk

[3] WHO. Environmental Noise Guidelines for the European Region. Copenhagen: WHO Europe; 2018

[4] Ross Ethier C, Simmons CA. Introductory Biomechanics: From Cells to Organisms. Cambridge: Cambridge University Press; 2007

[5] Motro R. Tensegrity: Structural Systems for the Future. London: Hermes Science Publishing; 2003

[6] WikiCommons. Needle Tower by Kenneth Snelson in Hirshhorn Sculpture Garden. Photo by Ben Stephenson. https://commons.wikimedia.org/wiki/File:Converging_Pattern.jpg

[7] WikiCommons. Tensegrity Icosahedron. Design first exhibited by Buckminster Fuller in 1949. Line Drawing by Bob Burkhardt. https://commons.wikimedia.org/wiki/File:Tensegrity_Icosahedron.png

[8] Ingber DE. The architecture of life. Scientific American. 1998;**278**:48-57

[9] Ingber DE. Mechanobiology and diseases of mechanotransduction. Annals of Medicine. 2003;**35**:1-14

[10] Paoletti S. The Fasciae: Anatomy, Dysfunction and Treatment. English edition. Seattle: Eastland Press; 2006

[11] Lindsay M, Robertson C. Fascia: Clinical Applications for Health and Human Performance. Delmar: Clifton Park; 2008

[12] Mohr GC, Cole JN, Guild E, von Gierke HE. Effects of low-frequency and infrasonic noise on man. Aerospace Medicine. 1965;**36**:817-824

[13] Svigovyi VI, Kuklina OI. State of the hemolymph circulatory bed of the conjunctiva as affected by infrasound. Gigiena Truda i Professional'nye Zabolevaniya. 1985;**6**:51-52. [Article in Russian]

[14] Kosacheva TI, Svidovyi VI, Alekseev VN, Kovalenko VI. Influence of noise and infrasound on the vision organs. Meditsina Truda i Promyshlennaia Ekologiia. 2001;(6):34-38. [Article in Russian]

[15] Van Zeller P, Tavares C, Mackay Freitas A, Oliveira A, Castelo Branco NAA. Fluorangiographic study of the ocular fundus in systemic vibration disease. Revista Portuguesa de Medicina Militar. 1991;**39**:67-70. ISSN 0482-7171 [Article in Portuguese]

[16] Qiu P, Zhang Z, Jiang Y, Gou Q, Wang B, Gou L, et al. Effect of infrasound on ultrastructure and permeability of rat's blood-retinal barrier. Zhonghua Yan Ke Za Zhi. 2002;**38**:499-501. [Article in Chinese]

[17] Fonseca J, Martins dos Santos J, Oliveira P, Laranjeira N, Águas A, Castelo Branco NAA. Noise-induced gastric lesions: A light and electron microscope study of the rat gastric wall exposed to low frequency noise. Arquivos de Gastroenterologia (Brazil). 2012;**49**:82-88

[18] Kim CY, Ryu JS, Hong SS. Effect of aircraft noise on gastric function. Yonsei Medical Journal. 1968;**9**:149-154

[19] Cohen A. The influence of a company hearing conservation program on extra-auditory problems in workers. Journal of Safety Research. 1976;**8**:146-162

[20] Castelo Branco NAA. The clinical stages of vibroacoustic disease. Aviation, Space and Environmental Medicine. 1999;**70**(Suppl):A32-A39

[21] Nekhoroshev AS, Glinchikov VV. Morphological research on the liver structures of experimental animals under the action of infrasound. Aviakosmicheskaia i Ekologicheskaia Meditsina. 1992;**26**:56-59. [Article in Russian]

[22] Ponomarkov VI, Tysik A, Kudryavtseva VI. Biological action of intense wide-band noise on animals. Problems of Space Biology. NASA TT F-529. 1969;7:307-309

[23] Svidovyi VI, Glinchikov VV. The effect of infrasound on pulmonary structure. Gigiena Truda i Professional'nye Zabolevaniya. 1987;**1**:34-37. [Article in Russian]

[24] Reis Ferreira JM, Monteiro MB, Tavares F, Serrano I, Monteiro E, Mendes CP, et al. Involvement of central airways in vibroacoustic disease patients. Revista Portuguesa de Pneumologia. 2006;**12**:93-105. [Thomé Villar/Boehringer Ingelheim Award]

[25] Fung YC. Biomechanics: Mechanical Properties of Living Tissues. New York: Springer-Verlag; 1993

[26] Shewry PR, Tatham AS, Bailey A, editors. Elastomeric Proteins. Cambridge, UK: Cambridge University Press; 2003

[27] Grande NR, Águas AP, Sousa Pereira A, Monteiro E, Castelo Branco NAA. Morphological changes in the rat lung parenchyma exposed to low frequency noise. Aviation, Space and Environmental Medicine. 1999;**70**(Suppl):A70-A77

[28] Castelo Branco NAA, Alves-Pereira M, Martins dos Santos J, Monteiro E. SEM and TEM study of rat respiratory epithelia exposed to low frequency noise. In: Mendez-Vilas A, editor. Science and Technology Education in Microscopy: An Overview. Vol. II. Badajoz (Spain): Formatex; 2003. pp. 505-533. ISBN: 84-607-6699-3

[29] Castelo Branco NAA, Gomes-Ferreira P, Monteiro E, Costa e Silva A, Reis Ferreira JM, Alves-Pereira M. Respiratory epithelia in Wistar rats after 48 hours of continuous exposure to low frequency noise. Revista Portuguesa de Pneumologia. 2003;**IX**:473-479

[30] Castelo Branco NAA, Monteiro E, Costa e Silva A, Reis Ferreira JM, Alves-Pereira M. Respiratory epithelia in Wistar rats born in low frequency noise plus varying amount of additional exposure. Revista Portuguesa de Pneumologia. 2003;**IX**:481-492

[31] Reis Ferreira JM, Couto AR, Jalles-Tavares N, Castelo Branco MSN, Castelo Branco NAA. Airway flow limitation in patients with vibroacoustic disease. Aviation, Space and Environmental Medicine. 1999;**70**(Suppl):A63-A69

[32] Castelo Branco NAA. A unique case of vibroacoustic disease. A tribute to an extraordinary patient. Aviation, Space and Environmental Medicine. 1999;**70**(Suppl):A27-A31

[33] Castelo Branco NAA, Águas AP, Sousa Pereira A, Monteiro E, Fragata JIG, Tavares F, et al. The human pericardium in vibroacoustic disease. Aviation, Space and Environmental Medicine. 1999;**70**(Suppl):A54-A62

[34] Castelo Branco NAA. Fragata JI, Martins AP, Monteiro E, Alves-Pereira M. The pericardium in vibroacoustic disease. I—Morphological features. In: Proceedings of the 12th International Congress on Sound & Vibration (ICSV12); Lisbon, 11-14 July 2005; Lisbon. Red Hook (NY): Curran Associates; 2005. pp. 1363-1371. ISBN: 978-1-62748-149-6

[35] Castelo Branco NAA. Fragata JI, Marques MC, Monteiro E, Alves-Pereira M. The pericardium in vibroacoustic disease. II. Cellular death pathways. In: Proceedings of the 12th International Congress on Sound & Vibration (ICSV12); Lisbon, 11-14 July 2005; Lisbon. Red Hook (NY): Curran Associates; 2005. pp. 1380-1387. ISBN: 978-1-62748-149-6

[36] Alves-Pereira M, Fragata JI, Monteiro E, Sousa Silva D, Castelo Branco NAA. The pericardium in vibroacoustic disease. III—A new structure. In: Proceedings of the 12th International Congress on Sound & Vibration (ICSV12); Lisbon, 11-14 July 2005. Red Hook (NY): Curran Associates; 2005. pp. 1372-1379. ISBN: 978-1-62748-149-6

[37] Marciniak W, Rodriguez E, Olsowska K, Botvin I, Araujo A, Pais F, et al. Echocardiography in 485 aeronautical workers exposed to different noise environments. Aviation, Space and Environmental Medicine. 1999;**70**(Suppl):A46-A53

[38] Araujo A, Pais F, Lopo Tuna JMC, Alves-Pereira M, Castelo Branco NAA. Echocardiography in noise-exposed flight crew. In: Proceedings of Internoise 2001; The Hague, 27-30 Aug 2001; Reston (VA); INCE-USA. 2001. pp. 1007-1010. ISBN: 9080655422

[39] Torres R, Tirado G, Roman A, Ramirez R, Colon H, Araujo A, et al. Vibroacoustic disease induced by long-term exposure to sonic booms. In: Proceedings of Internoise 2001; The Hague, 27-30 Aug 2001; Reston (VA); INCE-USA. 2001. pp. 1095-1098. ISBN: 9080655422

[40] Foraster M, Eze IC, Schaffner E, Vienneau D, HÉritier H, Endes S, et al. Exposure to road, railway and aircraft noise and arterial stiffness in the SAPALDIA study: Annual average noise levels and temporal noise characteristics. Environmental Health Perspectives. 2017;**125**:097004. DOI: 10.1289/EHP1136

[41] Alexeev SV, Glinchikov VV, Usenko VR. Infrasound induced myocardial ischemia in rats. Gigiena Truda i Professional'nye Zabolevaniya. 1983;**8**:34-38. [Article in Russian]

[42] Pei Z, Sang H, Li R, Xiao P, He J, Zhuang Z, et al. Infrasound-induced hemodynamics, ultrastructure, and molecular changes in the rat myocardium. Environmental Toxicology. 2007;**22**:169-175. DOI: 10.1002/tox.20244

[43] Pei Z, Zhuang Z, Xiao P, Chen J, Sang H, Ren J, et al. Influence of infrasound exposure on the whole L-type calcium currents in rat ventricular myocytes. Cardiovascular Toxicology. 2009;**9**:70-77. DOI: 10.1007/s12012-009-9037-3

[44] Antunes E, Oliveira P, Borrecho G, Oliveira MJR, Brito J, Águas A, et al. Myocardial fibrosis in rats exposed to low frequency noise. Acta Cardiologica. 2013;**68**:241-245

[45] Antunes E, Oliveira P, Oliveira MJR, Brito J, Águas A, Martins dos Santos J. Histomorphometric evaluation of the coronary arterial vessels in rats submitted to industrial noise. Acta Cardiologica. 2013;**68**:285-289

[46] Lousinha A, Oliveira MJ, Borrecho G, Brito J, Oliveira P, Oliveira de Carvalho A, et al. Infrasound induces coronary perivascular fibrosis in rats. Cardiovascular Pathology. 2018;**37**:39-44

[47] Jansen JA, van Veen AA, Bosch AA, van der Nagel R, Vos MA, Bakker JM, et al. Arrhythmia vulnerability of aged haploinsufficient Cx43 mice is determinant by heterogeneous downregulation of Cx43 combined

with increased fibrosis. Circulation. 2008;**118**:S494

[48] Antunes E, Borrecho G, Oliveira P, Brito J, Águas A, Martins dos Santos J. Immunohistochemical evaluation of cardiac connexin43 in rats exposed to low-frequency noise. International Journal of Clinical and Experimental Pathology. 2013;**6**:1874-1879

[49] Yuan H, Long H, Liu J, Qu L, Chen J, Mou X. Effects of infrasound on hippocampus-dependent learning and memory in rats and some underlying mechanisms. Environmental Toxicology and Pharmacology. 2009;**28**:243-247. DOI: 10.1016/j.etap.2009.04.011

[50] Mou X, Chen J, Li L, Jia KY, Qiu JY. Expression and distribution of glial fibrillary acidic protein in the brain of the mouse exposed to infrasound. Chinese Journal of Physical Medicine and Rehabilitation. 2001;**2**:76-78

[51] Du F, Yin L, Shi M, Cheng H, Xu X, Liu Z, et al. Involvement of microglial cells in infrasonic noise-induced stress via upregulated expression of corticotrophin releasing hormone type 1 receptor. Neuroscience. 2010;**167**:909-919. DOI: 10.1016/j.neuroscience.2010.02.060

[52] Kritas SK, Saggii A, Cerulli G, Caraffa A, Antinolfi P, Pantalone A, et al. Corticotropin-releasing hormone, microglia and mental disorders. International Journal of Immunopathology and Pharmacology. 2014;(2):163-167

[53] Shi M, Du F, Liu Y, Li L, Cai J, Zhang GF, et al. Glial cell-expressed mechanosensitive channel TRPV4 mediates infrasound-induced neuronal impairment. Acta Neuropathologica. 2013;**126**:725-739. DOI: 10.1007/s00401-013-1166-x

[54] Jiang S, Wang YQ, Xu CF, Li YN, Guo R, Li L. Involvement of connexin43 in the infrasonic noise-induced glutamate release by cultured astrocytes. Neurochemical Research. 2014;**39**:833-842. DOI: 10.1007/s11064-014-1277-3

[55] Ma L, He H, Liu X, Zhang G, Li L, Yan S, et al. Involvement of cannabinoid receptors in infrasonic noise-induced neuronal impairment. Acta Biochimica et Biophysica Sinica (Shanghai). 2015;(8):647-653. DOI: 10.1093/abbs/gmv049

[56] Shi YJ, Shi M, Xiao LJ, Li L, Zou LH, Li CY, et al. Inhibitive effects of FGF2/FGFR1 pathway on astrocyte-mediated inflammation in vivo and in vitro after infrasound exposure. Frontiers in Neuroscience. 2018;**12**:582. DOI: 10.3389/fnins.2018.00582

[57] Zhang MY, Chen C, Xie XJ, Xu SL, Guo GZ, Wang J. Damage to hippocampus of rats after being exposed to infrasound. Biomedical and Environmental Sciences. 2016;**29**:435-442. DOI: 10.3967/bes2016.056

Chapter 6

Acoustics from Interior Designer Perspective

Naglaa Sami AbdelAziz Mahmoud

Abstract

Should we consider the acoustics as engineering science or as architectural elements or as interior design applications? The main purpose of this chapter will focus on the differences between the three aspects of the acoustics with a special focus on the interior acoustical design. The arguments that favor the acoustic in each field are many. This chapter will clarify, strengthen, and explore the importance that has the acoustic study for the interior designing layout. From the historical eras where only material schemes were used, the acoustical treatment reached a great achievement. Nowadays, electrical, acoustical devices took place in different situations where only the materials could solve the acoustic needs. The considerations of using the electrical-acoustical devices remain under request, while their usage, only in specific conditions, a topic to highlight in this chapter.

Keywords: acoustics for interiors, soundscape, environmental acoustics

1. Introduction to acoustics

Acoustics is a multidisciplinary field where many specializations cohere. The more the knowledge advances, the more the subfield growths and thus the belonging matter struggle. Acoustics comprehends a numerous profession such as architecture, engineering, physics, electronics, speech communication, mechanical engineering, medicine and speech communication, music and arts, and oceanography. Acoustics is a title, which today needs to be more complex as terminology, as all of the mentioned professions consider it as their own.

Acoustics is associated until a near date to the engineering. However, nowadays, many titles raised to identify the stipulation of the specific field better.

The soundscape is the acoustics that deals with the unconscious effects of sounds and the surrounding sound background having precise psychological effects.

Acoustical environment is the interior design application of the sound behavior within the interior and the different physical reactions on the human functions. Acoustical oceanography is the study of the underwater sound. Acoustical engineering, like science, belonged to the engineering; then by the advancement of information, it splits to mechanical waves, where the science of sound and vibration across the technology exists. Acoustical engineering, as well, has a direct relationship with the sound control through electrical devices.

Interior acoustical design cooperates some of the previous elucidated aspects, the soundscape, the acoustical environment, and the electrical sound systems, where all are integrated into the interior spaces based on their functions.

IntechOpen

2. Acoustics and related field of specializations

When and how does humanity recognize acoustics?

The applications of the acoustics start from the historical eras for a diversity of purposes. From the prehistorical eras to the modern decades, especially with the emerging of the science and the recent innovations, the science of sound focused on informing, healing, and amusing. The main purpose of acoustics in design is the hearing of the required sounds. The diversity of disciplines, which deal with the acoustics, arises through some fields. R. Bruce Lindsay created "Lindsay's Wheel of Acoustics" (**Figure 1**) [1]. This wheel shows the fields of acoustics starting with the four broad fields of earth sciences, engineering, life sciences, and the arts. The outer circle lists the various broad disciplines one may study to prepare for a career in acoustics. The inner circle lists the fields within acoustics to which many fields naturally lead. The highlighted area shows the room acoustics, or the interior acoustical design, as well as all the related fields affecting to some extent the interior acoustical design clarified within this chapter.

2.1 Acoustics in engineering

Acoustic engineering is the branch of engineering dealing with sound and vibration, on its physical aspects. It is the application of the science of sound and vibration, in technology, typically concerned with the design, analysis, and control of sound. It also covers additional uses of sound, from the use of ultrasound in medicine to the programming of digital sound synthesizers. Acoustics, like engineering,

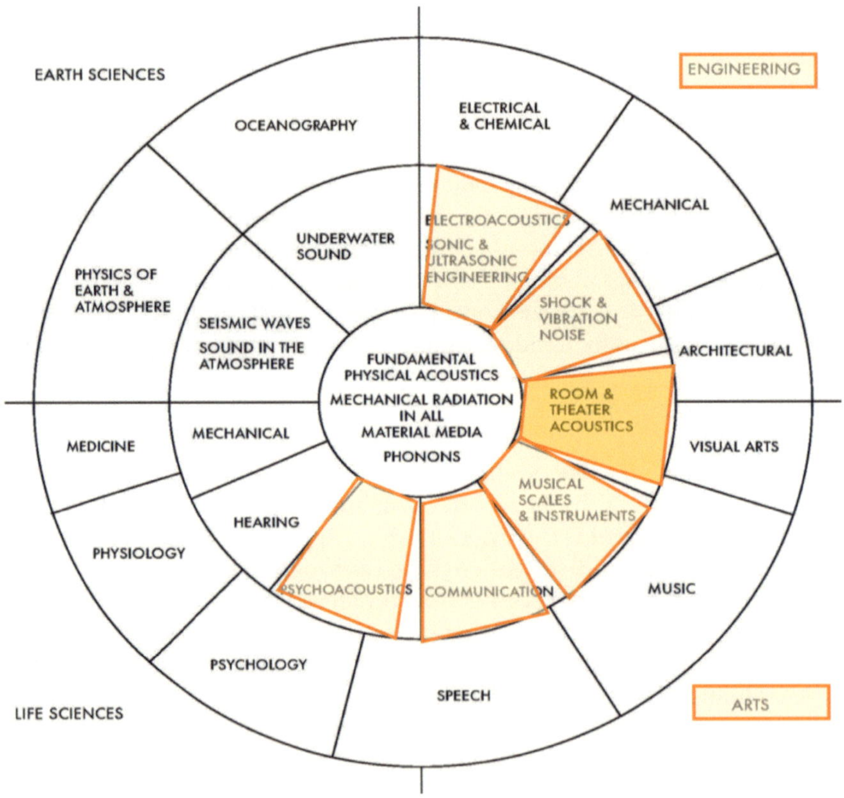

Figure 1.
Lindsay's wheel of acoustics.

deals with the electroacoustic elements, which are the electrical sound systems (microphones, headphones, loudspeakers, sound systems, sound masking, noise control, recording, and reproduction of sound). Acoustical engineer interprets with the sound from the physics perception. However, this side of the acoustics does affect the interior, as part of the whole design.

2.2 Acoustical design in the historical eras

A long time before the recent advancements and technologies, the different civilizations managed the available science they acquired from their previous ancestors and developed their daily life needs. By surfing among the different historical eras, we will discover that they had many solutions to solve their acoustical needs. Some of these solutions were, as per nowadays rules, architectural, and others were interiors!

From the prehistorical to the modern eras, diverse interpretation were used to enhance the sound level and the sound projection within interior spaces for the benefits of the event. Most of these events were educational, and their purpose was informing the communities with specific attained knowledge.

2.2.1 Acoustics in the prehistorical era

At the beginning of civilizations, the hunters, once back to the village, reproduce the hunting scenes for their young generation to teach them the procedures and tactics. They start to realize that the circular layout is the best to make others see and hear the message. They meet in a circular layout; either they are in the open air or their tents (**Figure 2**).

2.2.2 Acoustics in the Greek and Roman eras

By the Greek and Roman eras, the communities' fascination was the philosophy and the poetry. They build the amphitheaters, where speech and music could act, to communicate their knowledge. The shape of these buildings was to enhance the sound dissemination within the space (**Figure 3**). They used the open air to add extra reinforcement to the sound projection. At the same time, the open air works as a reflection to amplify their speech sound, in addition to the "horseshoe" plan layout and the level in the amphitheater style of interior layout, which enhances the acoustic performance of the place [4]. The steps of the audience area act as amplifier additionally to the rule of "Seeing ensures Hearing." As much, the person sees the stage as much the sound reaches the person without obstacles, although it

Figure 2.
Hunting scenes in the prehistorical era, in tents and open air [2].

Figure 3.
The amphitheater (the Greek Odeon of Herodes Atticus in Greece on the right and the Roman Theater of Mérida in Spain on the left) [3].

needs some reinforcements. The high edifice was using the background of the stage performance as backstage services spaces, stage set design, and main reflection elements to strengthen the sound on the stage.

2.2.3 Acoustics in the Islamic eras

The Islamic civilizations lasted for many eras and were a mix of religious reverence, science, and art lifestyle. They used the sciences as a base to all their inventions and creations under the spiritual beliefs. The great palace of Ālī Qāpū relieves new challenge of acoustical behavior and treatments. Built in the early Safavid style was the six-story Ceremony Hall (**Figure 4**). The muqarnas and cutout designs function to decrease reflections and act as a sound diffuser for the enhancement of the religious songs without echo, in beauty and functionality. Due to the distinguished advancements of mathematics, especially the algebra and physics, the results are magnificent [5].

2.3 Acoustics in architecture

Architectural acoustic design starts from the acoustical engineering. The connection between the physics and their applications when Wallace Clement Sabine, father of the modern building acoustics, gets the attention of the sound in the different lecture halls in Harvard where he studied physics science. Absorption formula title—Sabine's Formula—honoring his hard studies and experiments to reach today science the Acoustical Design in Buildings. The acoustics of buildings face several obstacles: the modern design trends, the type of materials, the mechanical systems in the buildings, and the surrounding level of noise.

The modern trends in design open up the spaces, no matter the functions and the privacy required for specific zones. Accordingly, the masking noise between the interface and the user concentration disappears.

Figure 4.
The six-story ceremony hall in Ālī Qāpū palace in Isfahan, Iran.

The building materials become economic in terms of structural properties in favor of space and cost saving. Such products create extra load on the noise production and prevent a safe environment for the users.

Mechanical systems in the buildings are necessary with today's climate. The air-conditioning systems, the elevators, and the ventilation are all components that contribute to the noise pattern of the buildings.

The level of noise's pollution created by today's lifestyle that is transmitted to the building add additional problems and multiply to a disastrous extend the noise pollution of our environment.

These problems are the main concern of the architectural acousticians.

2.4 Acoustics in interior design

Interior design, as specialization, dates to 1890. Elsie De Wolf developed the Colony Club in the US, and from then, it becomes a profession. The architect—the worker' supervisor in Latin—was in charge of all the building details, including the interior. Well, the interior design is the profession that cares, protects, and sustains the human life in a healthy, green, and safe environment. The interior acoustical intermediate that is appropriate to the function, reaching the audience in proper level to success the communication, without disturbing or even harming the users hearing, is a prosperous acoustical design. The classification of the interior spaces could lead to a better understanding of the role of the acoustical design to succeed in the functionality of the space. The interior acoustical design differs from the spaces where speech intelligibility is necessary to the spaces where quietness is vital, to spaces where the music needs enhancements, to spaces of private communications, and to spaces where public announcements take places. Each of these categories will require a specific combination of materiality schemes that balance between the physical properties and their quantities to provide an adequate acoustical environment [6].

2.4.1 Spaces for speech intelligibility

Speech clearness is vital in a specific type of functions. Without speech intelligibility, classes' success could fail. Teacher explanations, court sessions, and board meetings are functional, practical examples where speech is the key role of their success. Large institutions need clear discussions to make efficient decisions. In the court, when the speech is not clear, many judicial problems could take part from innocently. Such spaces need clear speech with no reflection nor echoes. In such functions, the number of absorber materials should increase in favor of the reflectors. The absorption material scheme when exceed, will produce a harmful background and thus will require more effort from the sources to enable the audience to hear. Therefore the introduction of some diffusers is necessary to balance the acoustics performance (**Figure 5**).

Figure 5.
Examples of classroom, boardroom, and court room where the speech intelligibility needs many absorbers.

2.4.2 Spaces for quietness needs

Sick peoples are the top users of quiet spaces where they take their recovering period. Museums are the place for specific levels of emotional entertainment or educational session; both of them are a relatively individual requirement and need high quietness levels. Classrooms are the spaces where instructors and students met for the educational processes environment. All these spaces (**Figure 6**) do require the best level of quietness background to achieve their specific functions. Hospitals, museums, and classrooms need a quiet environment. Such acoustical needs require more than the interior material scheme, it needs the isolation treatments to reach the quietness level.

2.4.3 Spaces for music enhancements

Recording studios, musical halls, and theaters are the locations of musical enhancement functions. Each of these areas needs different acoustical treatments to reach a vital level of sound. Studios need absorption layout to remove any reflection possible. While the musical halls, where live music takes place, requisite a mixture between the reflection and diffusing materials to reach the life feeling of sound without echoes nor masking of sound (**Figure 7**). Theaters are relatively difficult as they could host drama performance and musical performance.

2.4.4 Spaces for privacy needs

Private discussion occurs in human resources offices, in a medical clinic, in residential spaces, and police stations. These environments need high absorption

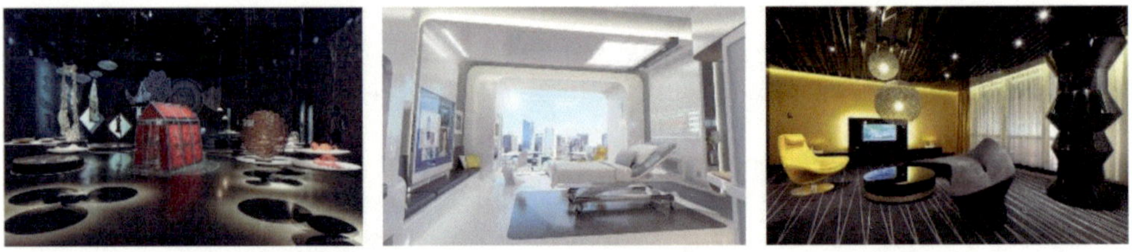

Figure 6.
Examples of museum, hospital, and living room where the quietness needs much absorbers and isolations.

Figure 7.
Examples of concert hall where the music needs to be enhanced through the reflectors and the diffusers.

DOI: http://dx.doi.org/10.5772/intechopen.84167

Figure 8.
Public announcement (main corridor in a mall where absorber patterns in the ceiling on the right and principal hall in airport where absorber baffles on the left).

quality to promote the privacy necessary. One of the human feelings is to feel shame from exposing private issues. Respecting and protecting these feelings are mandatory in designing such spaces. Such spaces could exist in buildings where other acoustical requirements are in question. Example of Doctor Room that needs privacy, during the clinic, and called for the quiet environment.

2.4.5 Spaces for public announcements

Airport, public malls, and governmental spaces for individual services are all spaces where public announcement occurs. The announcement of the flight number and timing requires a clearness in the speech to understand the announced information. Otherwise, the disruption and the distraction are the results. Similar results could happen in the public spaces as malls or governmental facilities (**Figure 8**). Acoustic design plays a different role in creating a masking background to clear the pronounced data.

3. Acoustical interior design elements

Interior acoustics support the well-being of the users. The design of the spaces that require a sound quality needs specific elements to embolden the acoustical quality. The interior acoustical elements involve and interlace with other acoustical fields. These elements start by the space layout, which should be architecturally created to the specific function to prevent any basic problems. Otherwise, the interior solutions could cover such deficiency. The selective materials are the essence of the interior acoustical design. The interior treatments, quality, and position play the main role in the excellence of the acoustical functionality. It is a pure interior design, although the architect and especially the architectural acoustician consider it, to a very recent date, architectural treatments. The acoustical sound systems come as the third element as it is the additional element to any acoustical interior, for specific targets. If the acoustical interior design reflects the user's function, the acoustical engineering solutions presented through the sound systems should have a specific objective.

The sound design, as a background, affects the users unconsciously. It could contribute to the psychological mood of the users. Human is productive while listening to his/her favorite music by 30%. The type of sound background—or the soundscape—contributes in the overall success of the interior functions. Slow music accelerates the blood circulation, therefore speeding up human functionality. On the other hand, the slow music depresses most of the time the human being. Soundscape, as a new psychological tool, bolsters in today interior design upgrading the overall interior purposes.

The interior acoustical design is a combination of these three important elements that create the interior spaces in excellence [7].

3.1 Interior acoustical layout

The architectural design affects deeply in the interior sound behavior. While the sound behavior follows the architectural layout, the architectural shape could lead to a successful acoustic or a harmful one. The architecture should support the acoustical design to enhance the functionality rather than create problems that need extra solutions. Domes, circular plan, parallel surfaces, and unproportioned spaces lead to several acoustical problems for the interior spaces. Solving these problems will over cost the interior, as the acoustical treatments are expensive as raw materials and as fire-resistant coatings [8].

3.1.1 Domes

Dome does reflect the sound to a specific point, creating a nonstop echo. Adding absorbers on the full surfaces of the dome will create a convenient sound layout for space. Suspending of a huge luminaire, designed for masking the echo, is a solution of cutting out the reflection but in condition to use an amount of absorbing materials within its design. Baffles are a different solution but, similar to the suspended luminaire, it will block the view of the dome and will create an additional different perception.

3.1.2 Circular plan

Circular plan is a challenge for the acoustical design perfection. Similar to some extent to the dome, straight surfaces of paneling would break the echo resulting from the circular perimeter. A mixture of balanced acoustical materiality scheme is the best solution, and the decision of the properties of the materials relates to the functional acoustical study of the specific interior [9].

3.1.3 Parallel surfaces

The interior surfaces should not have parallel surfaces, either as peripheral surfaces or as ceiling and flooring opposite to each other. The interior acoustical designer needs to break these parallelisms. The creation of acoustical interior treatments, diverse in quantities and properties, will solve this problem, although it is much expensive than having the spaces without such problems.

3.2 Interior acoustical treatments

The sound behavior in the interior spaces is the result of the interior treatments used. The quality and the properties of these materials share to some extent the success of the sound propagation within the space. The sound could be reflected, absorbed, diffused, or transmitted depending on the material physics (**Figure 9**). Materials used in the acoustical design are expensive as their majority are natural materials. Additionally, they need exclusive fire-resistant coating.

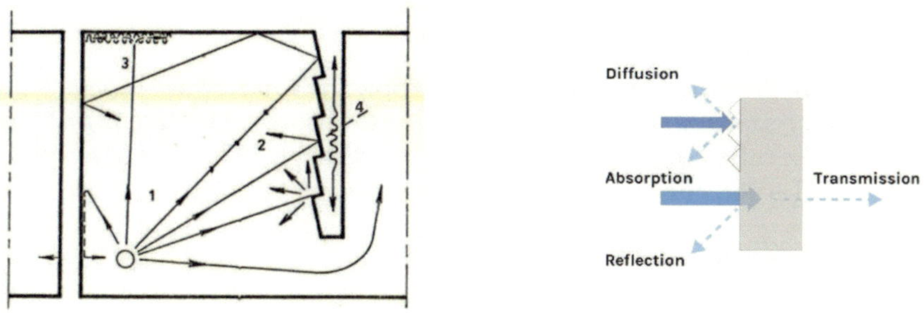

Figure 9.
Sound behavior in the enclosed spaces ((1) sound wave projected from the source, (2) sound wave reflected on the surface, (3) sound wave absorbed, and (4) sound wave transmitted).

3.2.1 Reflectors

Reflectors are hard solid materials that reflect the sound following the rule of reflection—angle of projection equals to angle of reflection but in the opposite side. Best reflectors are the ones made from natural hardwood with thickness limits of 5–15 cm (**Figure 10**). Oak, beech, mahogany, maple, walnut, and pine are all good selection. Ash wood is one of the best as it can easily bend which allows better coverage in favor of materiality limitations. Convex surfaces will scatter the sound waves allowing natural reinforcement for the unamplified voice without electrical sound systems while reducing the material surfaces.

3.2.2 Absorbers

Opposite to the reflectors, absorbers are soft, porous materials (**Figure 11**). Used to name the foams with different densities and thicknesses, a sandwich panel composed from a finished layer of fabric or soft natural perforated or slotted wood (with different perforation percentages), and the inside might be one or more of the following materials: rock wool, glass wool, or foam, based on its absorption coefficient "alpha". Heavy curtains exceed its dimensions seven times the length to work as absorbers.

Figure 10.
Reflectors appearance in the interiors on the left and the difference of dimensions between the flat reflectors in comparison to the convex reflectors on the right.

Figure 11.
From left, perforated and slotted softwood, sandwich panels (fabrics coverage with glass wool, rock wool, or foam), and different densities of foam.

3.2.3 Diffusers

Diffusers are the type of treatments that deploy the sound wave in a way to keep it alive without reflecting it to a specific spot nor to absorb it (**Figure 12**). Diffusers are similar to the reflector properties **but** in altered shapes. They are hard solid materials full of angles and curves to diffuse the sound waves within the interior spaces.

3.2.4 Isolation materials

The isolation materials usually are used to prevent the transmission of the sound waves from space to another. Spaces could be enclosed or open, but the essential is to block their flows and thus avoid the noise transmissions. Isolation materials consist of high-density absorbing materials in addition to reflectors in a composition that supports, scientifically and structurally, the sound blockage. However, they are part of the structure of the building layout, yet the acoustical interior may interface to solve specific noise transmission and therefore be part of the design. Building materials are the rescue to solve such situation in the interior acoustical design phases.

3.3 Interior acoustical sound system

Electrical sound systems for the acoustical interior design are not useful, especially when the interior wall is acoustically designed. It has dedicated purposes when recording constituents play a role in the functional aspects. It is worth to mention that this section is an acoustical engineering field that serves the interior and its inhabitants directly. For the acoustical interior designer, the sound system exposure to three main questionings, the what, the why, and the how of the electrical sound system [10].

What are the sound system and its components? Why should we use the sound system? How to apply the sound system within ideal regulations?

3.3.1 Sound system components

What are the sound system and its components? The sound system encompasses three essential elements: the microphone, the loudspeakers, and the controlling devices (**Figure 13**). Microphones are diverse in shape, design, and specifications. Each serves a specific function. The self-switch, the controllable, and the standard normal are all options for the specific situation in the overall interior function and needs. Electrically spoken, the types of microphone differ based on the frequency response, internal circuitry, resistance to the moisture, diaphragm size, weight, and durability. Concerning the loudspeaker, it consists of main parts: the woofer for the low frequencies, twitter for the high frequencies, and the midrange. Finally, the control panel to amplify the sound based on the specific results is required [11].

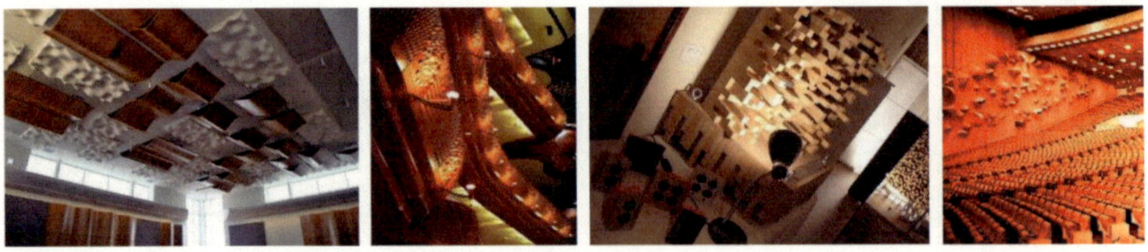

Figure 12.
Diffusers in different appearances of curves and angles, through the hard solid selection.

Figure 13.
Sound system components (microphone, loudspeakers, and control panel).

3.3.2 Sound system purposes

Why should we use the sound system? The sound behavior in the interior spaces is the result of the interior treatments used. The quality and the properties of these materials share to some extent the overall quality of the sound. Using the sound system is for recording the event (unrepeated or memorial event), for listening to recorded materials (like the cinema), and for the simultaneous translation. Except for these three purposes, the designer does not refer to the sound system since the interior which follows the standard setup will never necessitate such devices.

3.3.3 Sound system regulations

How to apply the sound system within ideal regulations? The sound system should follow strict regulations to ensure the quality level of the sound projected. Therefore, the strict regulations (**Figure 14**), once applied, will provide the best sound system results. The selection of sound system properties contributes to the overall design of these sound systems in the interior layout. The angles of the sound projections from the loudspeakers determine the numbers of speakers and their positions in the interiors.

The international meeting and event necessity are the simultaneous translation. In this case, the translation booths should be available by numbers equal to the six

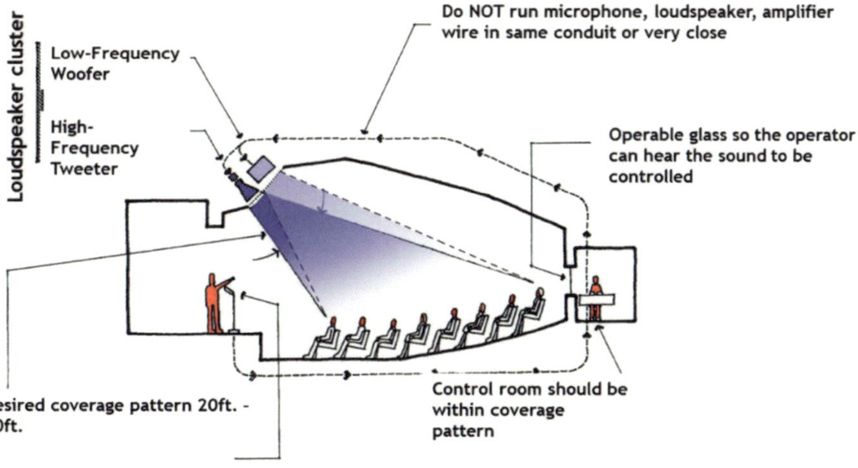

Figure 14.
Sound system regulations in a section layout [6].

official international languages. These booths should be enough in sizes to host, in each, two translators—each takes speech for 10 min. The visual field of these booths in the space must be clear to the main panel of discussion on the stage.

4. Interior acoustical design criterion

Interior acoustical design is a relatively new profession, yet the acoustics was one of the human concerns throughout history. By primitiveness, humans from the prehistorical dates, automatically, used their instinct to adjust their interaction fields in a way to reach the best paths of sound communications.

Lindsay's wheel of acoustics shows the selective area of the architectural interior design, but the related fields that share in some range their outcomes and make benefits to the interiors are many (highlighted in the wheel (**Figure 1**)).

Interior acoustical design is a field where mixes of discipline contribute to the sound excellence for the benefits of the users. Interior acoustical design involves the basics of physics as science relates the reflection, absorption data, the sound formula regarding the reverberation time, the physical measurements of the materials based on the coefficient of absorption (alpha $\bar{\alpha}$), the space volume, and the sound frequencies. All these inputs enabled the creation of the new interior acoustical wheel (**Figure 15**) [12].

Interior acoustical design implicates in parallel the artistic perception of the material selection, the material patterning, and the material color schemes. Therefore, the interior acoustical design is deliberated, as science and art, through the following points:

- The **classification of functional interior** permits the selection of the design paths regarding materiality schemes that support the interior function in favor of the users. The five different space functionality classifications are speech intelligibility, quiet interior, music enhancements, privacy needs, and public announcements.

- The **soundscape** plays a major role in the psychology scheme of the space as much it affects the users in unconscious ways.

- The interior acoustical space needs hard efforts to prevent the weakness of the **architectural layout** from the architectural perceptions, the domes, the

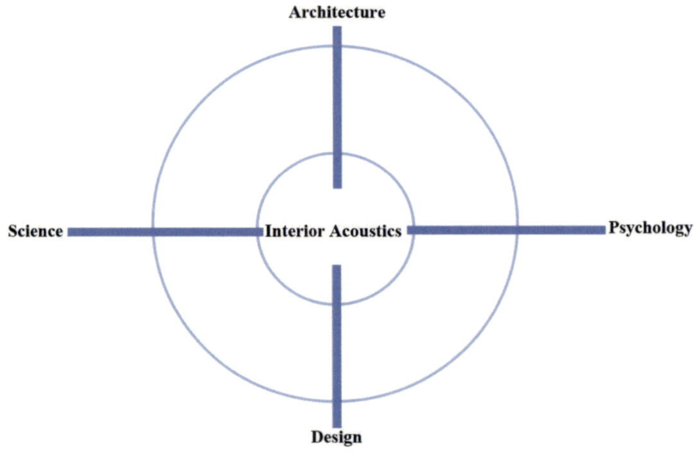

Figure 15.
Interior acoustical design wheel.

parallel surfaces, the unproportioned spaces, and the circular plan's to reach an appropriate acoustical interior design.

- The interior acoustical design—or **room acoustics**—depends on the quality of the **material properties** and their **positions** in the interior.
- The interior acoustical material classifications are reflectors, absorbers, and diffusers.
- **Reflectors** are hard solid materials, where the designer can use simple formulas to determine their position (**Figure 16**), where "A" is the projected ray of sound, "B" is the reflected ray of the sound, and "C" is the direct ray of the sound.
- **Absorbers** are soft porous materials, where the coefficient of absorption alpha ($\bar{\alpha}$) specifies their level of absorption. The material is absorber if $\bar{\alpha}$ = 0.6–1 (**Figure 17**). Absorbers are composite materials (sandwich panels).
- **Diffusers** are hard solid materials full of curves or hard edges (angles).
- Isolation materials are within the architect responsibility, yet the interior designer could refer to them when needed.
- The **sound system** is designed by the electrical engineers, yet it is part of the acoustical interior layout upon strict guideline of uses. The three functional parameters of the electrical sound systems are recording the events, listening

A+B-C≤12m for speech
A+B-C≤17m for music

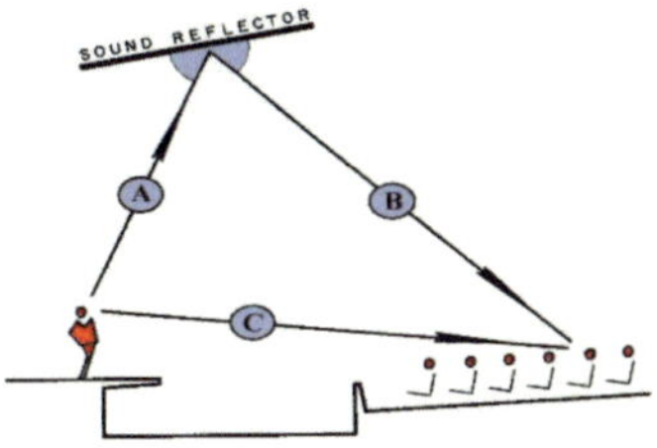

Figure 16.
Reflection position formulas [13].

Material	Frequency (Hz)*					
	125	250	500	1.000	2.000	4.000
Acoustic Panels	0.15	0.3	0.75	0.85	0.75	0.4
Brick	0.024	0.024	0.03	0.04	0.05	0.07
Carpet	0.05	0.1	0.2	0.25	0.3	0.35
Curtains	0.05	0.12	0.15	0.27	0.37	0.5
4" Rockwool	0.38	0.89	0.96	0.98	0.81	0.87
Wood Floor (Joists)	0.15	0.2	0.1	0.1	0.1	0.05
Glass	0.03	0.03	0.03	0.03	0.02	0.02
Seated Person	0.18	0.4	0.46	0.46	0.5	0.46
Plasterboard	0.3	0.3	0.1	0.1	0.04	0.02
Plywood on 2" Batten	0.35	0.25	0.2	0.15	0.05	0.05
¾" Wood Panel	0.1	0.11	0.1	0.08	0.08	0.11

*Note: A ($\bar{\alpha}$) coefficient of 1.0 means 100% Absorption, such as an open window, while 0.0 means 100% Reflection, such as glass. All figures are given for one square meter of materials.

Figure 17.
Alpha chart to standard frequencies.

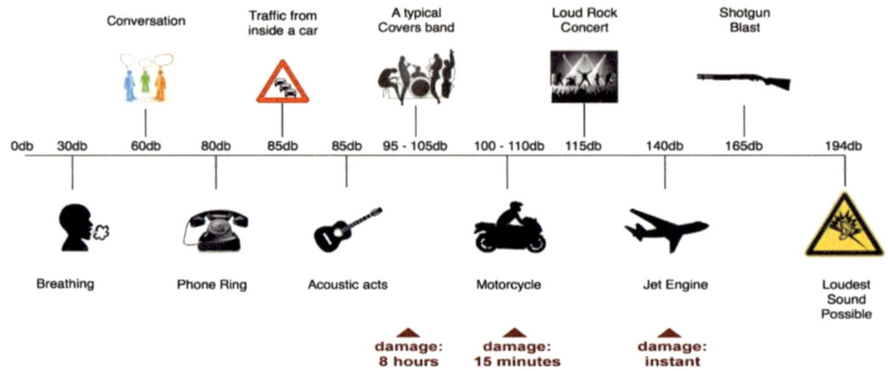

Figure 18.
Sound level chart to the normal daily life, where the hearing damages are indicated.

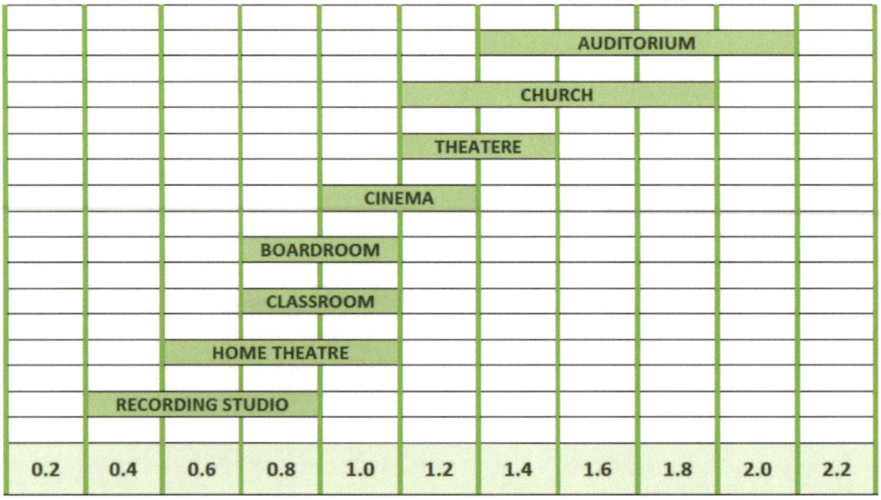

Figure 19.
Reverberation time (RT) measured in seconds for specific functions.

to recorded materials, and simultaneous translation. The sound system should follow the rigor layout for the benefits of the output sounds.

- The **sound level** (measured in decibel—"dB") (**Figure 18**) appropriate to the function is standard, and the interior acoustical designer needs to follow the mathematical steps (using Sabine or Earing Formulas) to reach the convenient reverberation time (RT) (**Figure 19**). These steps and formulas are part of the mechanical engineering field.

- The interior acoustical design is a combination of science, architecture, psychology, and design that contribute in interlacing paths to reach the sound excellence for the wellness of the human being.

5. Conclusion

The acoustics from interior designer perspective reach the end. The following indices could conclude the main features of discussion, already detailed within this chapter:

- Acoustics dates from the prehistorical eras, where the human considered it a vital component of the daily life, to sustain the communication.
- The uses of the architectural layout and the interior components were the tools to apply a good acoustical environment, using the science, within the historical periods.
- Interior acoustic design needs a clear classification of functional necessities to reach a successful synthesis that suits the end users of the spaces.
- The interior acoustical success relies on the quality and the locus of the materials used within the space upon stricken regulations.
- The material quality determines either it is a reflector, absorber, or diffuser.
- Reflectors are hard, rigid, and thick materials.
- Absorbers are soft and porous, and, usually, they are in the form of "sandwich panels."
- Diffusers are similar to the reflectors but never in plane surfaces, only if full of curves or angles.
- The coefficient of absorption (alpha $\bar{\alpha}$) factor permits to apportion the material scheme within the interior. Alpha differs upon the level frequencies. When overall alpha magnitude reaches the round of "1," it means the material is absorber, and when reaching the round of "zero," it means the material is reflector.
- Sound systems are used for only archetypical purposes and should follow the confine regulations.
- Soundscape is an element to consider effectively, as it affects the human psychological within the interior layout.

Acknowledgements

I would like to express my gratitude to my professor, my mentor, and my father Professor Sami Abdelaziz Mahmoud, who was the main reason of my professional life, especially in the field of lighting and acoustics.

Nomenclature

Reflectors	are hard solid materials to reflect the sound waves
Absorbers	are soft porous materials to absorb the exceed sound
Diffusers	are hard sold materials full of edges to scatter the sound
dB	decibel: measurement unit of the sound level. "deci" logarithm unit and "Bel" in honoring Alexander Graham Bel, the founder of the sound science
RT	Reverberation time is the lifetime of the sound until it becomes mute and measured by seconds
Hz	hertz, the measurement unit of the sound frequencies
Woofer	the part in any loudspeaker responsible of the low frequencies
Tweeter	the part in any loudspeaker responsible of the high frequencies

Author details

Naglaa Sami AbdelAziz Mahmoud
Interior Design Department, Ajman University, Ajman, UAE

*Address all correspondence to: n.abdelaziz@ajman.ac.ae

References

[1] Fields of Acoustics Chart [Internet]. 1929. Available from: https://exploresound.org/explore-sound-home/what-is-acoustics/fields-of-acoustics/ [Accessed: November 26, 2018]

[2] Prehistorical Gathering [Internet]. 2015. Available from: https://www.thinglink.com/scene/715694819694870528 [Accessed: November 26, 2018]

[3] Ancient Amphitheaters [Internet]. Available from: https://www.touropia.com/ancient-theatres-of-greek-roman-antiquity/ [Accessed: November 28, 2018]

[4] Fuchs H. Briefing: Did the ancient Greeks know acoustics better? Proceedings of the Institution of Civil Engineers - Engineering History and Heritage. 2009;**162**(4):175-178

[5] Azad H. Ali Qapu: Persian historical music room. Akutek. 2008;**30**(3):208-215

[6] Long M. Architectural Acoustics. USA: Academic Press; 2014. ISBN: 9780123982582

[7] Hayato S, Masayuki M, Hiroshi S, Megumi W. Relationship between listening difficulty and acoustical objective measures in reverberant sound fields. The Journal of the Acoustical Society of America. 2008;**123**(4):2087-2093

[8] Abdelaziz S. The effects of the acoustical designing on the architectural designing form. Vekoslav Pardyl. Prague: Czech Technical University; 1977

[9] Beckers B, Borgia N. The acoustic model of the Greek theatre. France: Urban Systems Engineering Department, Compiegne University of Technology; www.heliodon.net

[10] Edie OJ, Enoma EP, Igbogbo OS, Ezema IC, Ekhaese EN. A study of the effects of architectural forms on sound quality in church buildings. Nigerian Journal of Environmental Sciences and Technology. 2017;**1**(1):43-54

[11] Long M. Architectural Acoustics. Oxford/Kidlington: Elsevier; 2014

[12] Hamayon L. Reussir L'Acoustique d'un Batiment. Paris: LeMoniteur; 1996

[13] Leslie D. Environmental Acoustics. London: McGraw-Hill; 1972

Section 3

Nondestructive Testing (NDT), Non Linearity, Leakage

Chapter 7

Time Domain Analysis of Elastic Nonlinearity in Concrete Using Continuous Waves

Mourad Bentahar, Charfeddine Mechri, Paola Antonaci, Antonio Gliozzi and Marco Scalerandi

Abstract

Concrete and consolidated granular media in general exhibit a strong nonlinear hysteretic elastic behavior when excited by ultrasonic wave perturbations. Due to the sensitivity of their elastic properties to the small changes that can appear in their microstructure, the dynamic stress-strain relationship considered at low strains is affected by the presence of microcracks and hence the progression of damage. Tracking the nonlinear behavior can be made through the dependence on the excitation amplitude of the amplitude of higher order harmonics or of the resonance frequency of the sample. The present chapter shows a time domain analysis of elastic nonlinearity based on the break of the superposition principle when ultrasonic continuous waves are propagating in concrete samples. The latter, which can be of different microstructures (grain sizes, mortar, or polymer matrix), helps to understand the physical mechanisms involved in the different nonlinear elastic responses.

Keywords: concrete, damage progression, nonlinear ultrasound, nonlinear elasticity, microcracking, nondestructive testing

1. Introduction

Numerous experimental observations have shown that microinhomogeneous media manifest a "new" nonlinear elastic behavior different from the one explained by the classical theory of nonlinear elasticity proposed by Landau and applicable to homogeneous and crystalline media. The "new" properties, described in the framework of the nonlinear mesoscopic elasticity (NME) formalism, include most types of rocks, concrete, bones, damaged metals, and composites, which manifest the same macroscopic observations despite the existing differences in their microstructures and chemical constituents [1–8]. Indeed, all these materials share the characteristic of being complex with contacts, or microdefects (i.e., cracks), grain boundaries, dislocations, etc.

Observations on these materials concern quasi-static and dynamic acoustic experiments. In quasi-static tests, the stress-strain relationship is governed by an unusual multivalued function where the observed hysteretic loop contains small inner loops showing the presence of a memory effect [9]. In dynamic experiments with a propagating acoustic wave, the proportionality between input and output elastic waves is no longer valid. In such case, wave propagation (or even in standing wave conditions) is accompanied by the generation of amplitude dependent higher harmonics and sidebands [6, 10]. In addition, in resonance experiments, the increase in the dynamic perturbation creates a decrease in the elastic modulus of the propagating medium. This effect, which might be local or global, is observed through a decrease in the resonance frequency showing thus a softening in the elastic properties with an increase in damping around the excited resonance mode [11]. For relatively large excitation amplitudes, in the reversible regime, experiments show the presence of conditioning which means that the softening of elastic properties persists even when the excitation is switched off. In this case, at a given excitation amplitude, it takes seconds to minutes to stop the softening process and leads the medium into a new "equilibrium" state. The conditioning stops when the excitation is switched off. In this case, the medium needs minutes to days (depending on its state) to go back to its initial elastic state. This process is called relaxation and evolves in log-time [11–13].

So far in the literature, the aforementioned nonlinear effects have all been grouped into the same class in contrast with the classical nonlinearity well described by the Landau theory. However, a link has been postulated between the macroscopic nonlinear response and the microstructure of the different media [14]. Neutron scattering measurements have confirmed that the nonlinear behavior is localized in small regions close to discontinuities [15]. The observed regions might have different properties which makes the physical mechanisms described by the constitutive equation different. This makes the development of research around the existing relationship between the microscopic features and the different macroscopic observations interesting for basic research and for microcracks diagnosis in complex media.

Based on the above definitions, concrete as a consolidated granular medium is complex. Indeed, it exhibits a complicated nonlinear elastic behavior including hysteresis, harmonics generation, loss of reciprocity, etc. The complex structure of concrete can be seriously affected when damage is present. The latter, which can be of different origins, leads to important changes in the quasi-static (for advanced damage stages) and dynamic (for early damage stages) responses of concrete. Steel corrosion in reinforced concrete elements, mechanical stresses, thermal stresses, chemical attack by expansive agents, etc. have all negative effects on the concrete load carrying capacity, since they all lead to an increase in crack density and propagation. In this chapter, we propose to study the efficiency of a time domain analysis of elastic nonlinearity using continuous waves propagating in progressively damaged concrete samples.

2. Classical nonlinear theory

2.1 Stress-strain relations

In the approximation of small deformations, the free energy of an elastic system can be expanded as a power series with respect to the strain tensor. The general expression for the free energy of an isotropic body in the third approximation can be reduced by the convention of Einstein notation to

$$F = f_0 + C_{ijkl}\,\epsilon_{ij}\epsilon_{kl} + C_{ijklmn}\,\epsilon_{ij}\epsilon_{kl}\epsilon_{mn} \tag{1}$$

where f_0 is the constant, and C_{ijkl} and C_{ijklmn} are, respectively, the second and third order elastic tensors. For small strain amplitudes, the cubic term in the elastic energy can be neglected and the free energy reduces to

$$F = f_0 + C_{ijkl}\,\epsilon_{ij}\epsilon_{kl} \tag{2}$$

In the specific case of isotropic materials, the expression could be further reduced as

$$F = f_0 + \mu\,{\epsilon_{ij}}^2 + \frac{1}{2}\,\lambda\,{\epsilon_{ii}}^2 \tag{3}$$

where μ and λ are the Lamé coefficients.

At larger but still infinitesimal strains, the contribution of the third-order elastic term in the free energy expansion can no more be neglected. In such a case, the general expression for an isotropic medium becomes

$$F = f_0 + \mu\,{\epsilon_{ij}}^2 + \frac{1}{2}\,\lambda\,{\epsilon_{ii}}^2 + \frac{1}{3}\,A\,\epsilon_{ij}\,\epsilon_{ik}\epsilon_{kl} + B\,{\epsilon_{ij}}^2\epsilon_{ii} + \frac{1}{3}\,C\,{\epsilon_{ii}}^3 \tag{4}$$

where A, B, and C are the components of the third order elastic tensor for isotropic bodies. Consequently, the components of the stress tensor, obtained deriving the free energy with respect to strain are

$$\sigma_{ij} = \frac{\partial F}{\partial \epsilon_{ij}} = 2\mu\,\epsilon_{ij} + \frac{1}{2}\,\lambda\,\epsilon_{ii}\,\delta_{ij} + \frac{1}{3}\,A\,\epsilon_{ik}\epsilon_{kl} + 2\,B\,\epsilon_{ij}\,\delta_{ij} + C\,{\epsilon_{ii}}^2\delta_{ij} \tag{5}$$

Since strains are still infinitesimal, the corresponding elastic constants of the medium are

$$K_{ijkl} = \frac{\partial \sigma_{ij}}{\partial \epsilon_{kl}} \tag{6}$$

In the 1-D case, these equations reduce to the compressional stress σ

$$\sigma = (2\mu + \lambda)\,\epsilon + \left(\frac{1}{3}A + 3B + C\right)\epsilon^2 \tag{7}$$

The elastic modulus becomes

$$K = K_0(1 + \beta\epsilon) \tag{8}$$

K_0 is the Young modulus, and $\beta = \frac{\frac{1}{3}A + 3B + C}{2\mu + \lambda}$ is the quadratic nonlinear coefficient for longitudinal waves in isotropic media. Note that higher order terms of the elastic modulus can be obtained if we develop the free energy to the fourth order. In such a case, the modulus K would be $K = K_0(1 + \beta\epsilon + \delta\epsilon^2 + \ldots)$, where $\delta = \frac{3l + 2m}{2\mu + 2\lambda}$ is the cubic nonlinear coefficient and l, m are the 3rd order elastic constants (called Murnaghan constants).

2.2 Wave equation

In the 1-D case, the equation of motion corresponding to a longitudinal plane wave propagating in a quadratic nonlinear medium can be written as

$$\frac{\partial^2 u}{\partial x^2} - \frac{1}{C_L^2}\frac{\partial^2 u}{\partial t^2} = -2\beta\frac{\partial^2 u}{\partial x^2}\frac{\partial u}{\partial x} \tag{9}$$

where C_L is the wave speed in the linear medium. Eq. (9) can be rewritten as

$$\frac{\partial^2 u}{\partial t^2} = C_L^2 \frac{\partial^2 u}{\partial x^2}\left[1 + 2\beta\, \frac{\partial u}{\partial x}\right] = C^2 \frac{\partial^2 u}{\partial x^2} \tag{10}$$

where $C^2 = C_L^2\left[1 + 2\beta\, \frac{\partial u}{\partial x}\right]$ is the wave speed. Note that, when higher nonlinear parameters are considered, the wave speed becomes $C^2 = C_L^2\left[1 + 2\beta\, \frac{\partial u}{\partial x} + 3\delta\, \left(\frac{\partial u}{\partial x}\right)^2\right]$. The velocity then becomes strain dependent and could be affected by any change in the strain amplitude of the propagating wave as a consequence of the change in the elastic modulus.

$$\frac{C^2 - C_L^2}{C_L^2} = \beta\epsilon + \delta\epsilon^2 \tag{11}$$

As a consequence, in the case of a nondispersive medium, the same dependence could be observed and measured for the resonance frequency, w_r, the latter being proportional to the velocity

$$\frac{C^2 - C_L^2}{C_L^2} \propto \frac{w_r^2 - w_L^2}{w_l^2} \propto 2\frac{w_r - w_L}{w_L} \tag{12}$$

The strain dependence of velocity is resulting in a shift of the resonance frequency when strain (or stress) amplitude increases. Therefore, the relation between the resonance frequency shift and the strain amplitude could be written as

$$< \frac{w_r - w_L}{w_L} > \propto \beta < \epsilon(t) > + \delta < \epsilon^2(t) > \tag{13}$$

When the excitation is considered as a sinusoidal function, the frequency shift could be reasonably dependent on the higher order expansion term, where $\delta < \epsilon^2(t) > = \frac{1}{2}\, \delta\, \epsilon_{max}^2$, which makes $\frac{w_r - w_L}{w_L} \propto \frac{1}{2}\, \delta\, \epsilon_{max}^2$.

3. Nonclassical nonlinear wave propagation in solids

The above discussed theory revealed to be nonsufficient to describe nonlinear behavior of elastic waves in diverse solids (concrete, cracked metals or composites, rocks, etc.). Indeed, these materials revealed that they belong to the class of nonlinear mesoscopic elastic materials (NMEM), where several experimental observations are in contradiction with the classical Landau theory expectations. In quasi-static stress-strain experiments (performed on sandstone, for instance), the dependence of stress on strain was nonlinear, hysteretic and showed the presence of memory effect [16]. Such evolution cannot be predicted in the framework of classical nonlinear Landau theory. Therefore, we need to introduce an additional term into the definition of stress by taking into account new parameters such as the stress dependence on the sign of the strain rate.

In addition to the quasi-static behavior described above, other observations suggest that nonlinearity in NMEM materials should have a different origin from that of atomistic nonlinearity and should therefore be related to the material structure. Most of these experiments are showing an anomalous dynamic behavior in NMEM materials, that is, when a time dependent perturbation is applied. In the different work in the literature, there is an agreement about the fact that most undamaged materials such as intact aluminum, Plexiglas, and monocrystalline solids show only a very small nonlinear response related to strain at the atomic scale. In this range, their behavior is well described by the Landau theory. Evidence of such nonlinearity is only manifested at moderately high strain levels. Indeed, when these materials are microdamaged, their behavior at strain amplitudes $\varepsilon > 10^{-7}$ is more similar to that of Berea sandstone. Here, we should note that classical nonlinearity remains present, and as strain grows larger, its effect is hidden by stronger effects due to the presence of hysteresis. The following equation was therefore written as:

$$\frac{\partial^2 u}{\partial t^2} = C_0^2 \frac{\partial^2 u}{\partial x^2} \left(1 + \beta \frac{\partial u}{\partial x} + \delta \left(\frac{\partial u}{\partial x}\right)^2 + \dots\right) + H\left[\varepsilon, \text{sign}\left(\frac{\partial \varepsilon}{\partial t}\right)\right] \tag{14}$$

H is a function describing hysteretic nonlinearity, and $\frac{\partial \varepsilon}{\partial t}$ is the strain rate. Note that this function depends on the strain rate and on the strain history as well. However, we should point out that an analytic expression of the H function is still missing the reason for which only few discrete models have been proposed to reproduce and give some understanding of the experimental observations related to nonlinear mesoscopic elastic materials [17–19].

As it was discussed for the classical nonlinearity, the dynamic nonlinear response of NMEM may manifest itself in a variety of ways. Many indicators can therefore be defined to link the detected strain amplitude of the driving frequency to resonance shift, harmonics amplitude, break of the superposition principle, etc. In that case, additional indicators not existing for classical nonlinear materials can be introduced, in the sense that the observed effects on mesoscopic materials might be very different depending on the excitation duration (seconds, minutes, etc.). Therefore, two categories of experiments can be defined: fast dynamics, when the experiment lasts one or few periods of the perturbation, and slow dynamics when the response of the system is tracked on much longer time scale to observe conditioning and relaxation.

3.1 Fast dynamics

During a dynamic experiment, fast dynamic effects appear rapidly (the very early pico or nanoseconds are sufficient to observe the amplitude dependence) and could be observed using standing or transient waves.

Harmonic generation consists in exciting a sample with a source function and analyzing signals detected by receivers in the frequency domain by determining harmonic amplitudes via a Fourier analysis. Most experiments were conducted using compressional waves; however, some bending [20] and torsional wave measurements have been conducted as well [21].

The plot of the second and third harmonic amplitudes as a function of the strain amplitude of the fundamental remains a power law $y = ax^b$, as for classical nonlinearity. However, the calculation of the slope b of the same curve plotted in logarithmic scale provides a new quantitative information (i.e., in contradiction with what theoretically expected), certainly linked to the nature of the nonlinearity, where the exponent revealed to be the same for the second and third harmonics. In addition, for the same dynamic strain, the amplitude of the third harmonic (i.e., odd harmonics) is larger than the one corresponding to the second harmonic (i.e., even harmonics).

3.2 Nonlinear resonance frequency shift

Nonlinear effects can also be determined through the amplitude dependence of the resonance frequency through the technique named nonlinear resonant ultrasound spectroscopy (NRUS). The amplification provided by resonance makes

NRUS one of the most sensitive ways to observe nonlinear behavior, even at small dynamic strains ($\varepsilon \sim 10^{-8}$). In general, we can excite resonances by sweeping upward and downward around a given resonance frequency, and frequency sweeps are repeated at successively increasing amplitude over the same frequency interval. The frequency shift resulting from the different strain amplitudes helps learning about the nature of nonlinearity. Note that the dependence of the normalized frequency shift on the strain is a power law $y = ax^b$, where the exponent measured in different experiments revealed to be (b = 1) again different from the prediction of the classical Landau theory.

3.2.1 Slow dynamic effects

Slow dynamics is by far the most typical characteristic of the nonclassical NMEM. It refers to the logarithm dependence recovery of the elastic modulus to the original initial value (i.e., at rest) after being excited and therefore softened by a large amplitude strain. The log-time evolution of the elastic modulus resembles to a creep-like behavior observed in quasi-static experiments. However, it is important to note that slow dynamics is more likely a *new creep* behavior due to the fact that modulus is not following the symmetry of the strain. Furthermore, contrary to creep experiments, slow dynamics is a reversible and repeatable behavior, and observations on rocks and some damaged metals are performed at dynamic strain levels two or three orders of magnitude below those of a typical creep experiment. Slow dynamics includes two different scale mechanisms: conditioning and relaxation.

3.2.1.1 Conditioning

Conditioning or softening of the material takes place at dynamic strains corresponding to $\sim 10^{-6}$. Note that full conditioning could be obtained quite rapidly (few seconds to minutes), which seems to be long enough to allow neglecting its effects in fast dynamic experiments (much shorter time-scale). However, experiments show that most of conditioning occurs mostly instantaneously, which makes the coupling between fast dynamics and conditioning unavoidable, in the sense that the same wave propagating in a dynamic experiment is causing non-negligible self-conditioning. The elastic modulus decreases continuously during the dynamic excitation of NMEM until the material reaches a new equilibrium state where no more change takes place (see **Figure 1**). The amount of conditioning depends on the excitation time and amplitude [11, 22].

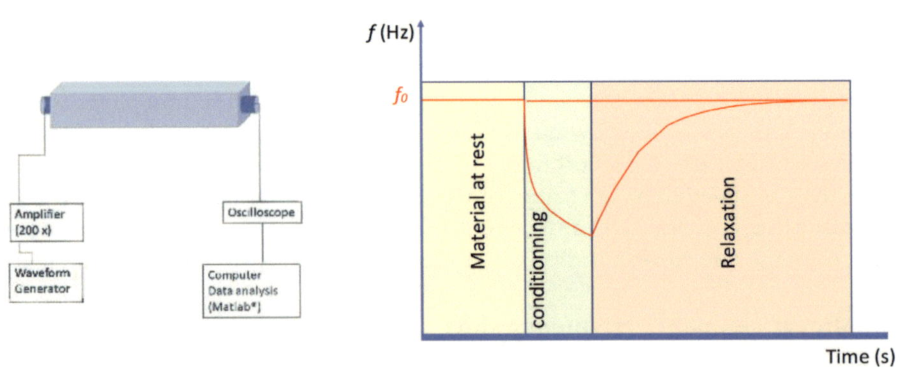

Figure 1.
Scheme of the experimental set-up (left); frequency (or elastic modulus) conditioning and relaxation evolution as a function of time (right).

3.2.1.2 Relaxation

Relaxation starts right after the full conditioning (see **Figure 1**). In practice, a first frequency sweep around a given resonance mode is performed at a small strain amplitude to verify that the material is relaxed (resonance frequency and damping remain unchanged in time). Then, the same excitation is applied at a very large strain level (equivalent to 10^{-6} or 10^{-5}) for few minutes. Afterward, successive sweeps are repeated at the lowest excitation amplitude (linear excitation $\sim 10^{-8}$) in order to probe relaxation around the excited resonance mode. Relaxation takes relatively large time and changes as the logarithm of time before the final recovery over minutes, hours, or days depending on the conditioning characteristics (strain amplitude, materials state, strain duration, etc.) [11, 22].

4. Experimental analysis

4.1 Experimental set up

Experiments are conducted generating ultrasonic signals through a waveform generator. Ultrasonic signals defined as monochromatic waves of amplitude A^{inp} and frequency w_0 are used to excite the sample under test not far from one of its compressional resonance mode (the fundamental in general)

$$u(t) = A_{inp} \cos(w_0 t) \tag{15}$$

The emitter transducer is glued to the sample using a linear coupling (phenyl salicylate, for instance). A second (identical) transducer is used to detect the response of the material under test, and it is connected to a digital oscilloscope for data acquisition. Signals are recorded in a short time window once stationary conditions are reached. In order to excite longitudinal modes, the transducers are put, in general, on opposite faces of the samples.

The experimental procedure starts by detecting the output signal at a very low excitation level A^{inp}. The latter is chosen the lowest possible in order to not have any change in the mechanical properties of the material under test. To verify this, we should have a good signal-to-noise ratio to generate output signals emerging from the noise level. In most of the presented experiments, the lowest amplitude was chosen as $A_0^{inp} = 5mV$.Under these conditions, the sample under test behaves almost linearly, and the recorded low amplitude response will be termed as "linear signal" $v_0(t)$. The linear signal measurement, performed without amplification in general, is followed by N acquisitions repeated increasing the amplitude of excitation up to a maximum level. The recorded N signals $v_i(t)$, $(i = 1, .., N)$ are recorded, each corresponding to an excitation amplitude A_i^{inp}.

4.2 The scaling subtraction method (SSM)

Under a dynamic excitation, the presence of nonlinearity can be detected through the validity of the superposition principle, which represents a requirement for a system to be linear. By considering a linear function F, if $v(t) = A\, v_0(t)$ is the input function, we have

$$u_A = F(Av_0) = AF(v_0) = Au_0 \tag{16}$$

F is the transfer function, u denotes the output signal, and A is the amplification factor. Here, u_0 is the response at the excitation amplitude v_0.

Consider an elastic wave propagating in a microdamaged medium. In such a case, one might expect that if the propagation excites the nonlinearity of the system, it will consequently break the superposition property. Therefore, if the exciting wave is generated at amplitude A_0, small enough so that nonlinearity of the medium is negligible, the system will behave linearly and its response is $u_0(t)$. At a larger excitation amplitude A, the response $u(t)$ of the same system is no longer equal to

$$u_{ref}(t) = \frac{A}{A_0} u_0(t) \tag{17}$$

which would be the response of that system if it remains linear even at large amplitudes. Therefore, the difference between the two responses can be taken as an indicator of nonlinearity. The nonlinear scaled subtracted signal $w(t)$, termed SSM signal, is introduced as (see **Figure 2**)

$$w(t) = u(t) - u_{ref}(t) \tag{18}$$

This time domain analysis of elastic nonlinearity, called scaling subtraction method (SSM), has proved to be sensitive to damage detection and easy to set [16].

From the quantitative point of view, if exciting signals are in the form of monochromatic continuous waves and measurements are taken in standing wave conditions, the SSM signal $w(t)$ is also a continuous wave. Thus, a parameter could be introduced either as the maximum or the "energy" of the nonlinear signal $w(t)$ as

$$\theta = \max(w(t)) \tag{19}$$

$$\theta' = 1/T \int_0^T w^2(t)dt \tag{20}$$

where T is the wave period. The parameter could then be shown as a function of excitation amplitude in order to highlight nonlinearity.

From the point of view of damage monitoring, for a given sample state, a quantitative nonlinear indicator must be defined. To this purpose, we observe that in materials exhibiting hysteresis, a power law holds in the form

$$\theta = ax^b \tag{21}$$

where x is the maximum of the output amplitude. Thus, experimental data could be fitted to derive the coefficient a and the parameter b, normally called slope, since

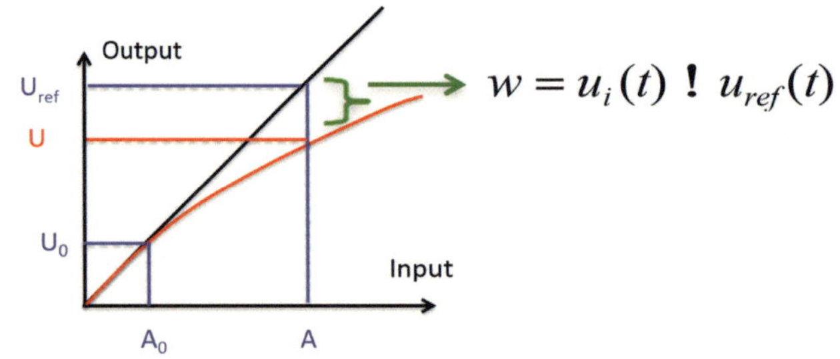

Figure 2.
Basic principle of SSM analysis.

in log-log scale, the above equation reduces to a straight line $20\ log_{10}\ \theta = a' + b20\ log_{10}\ x$.

As an alternative, the nonlinear indicator could be defined as the value of θ at a fixed value of x (excitation level).

5. Application to nonlinear characterization of concrete

Consolidated granular media and in particular concrete exhibit a strong nonlinear hysteretic elastic behavior when excited by ultrasonic wave perturbations [23–28]. The nonlinear behavior is strongly enhanced when the concrete is damaged [29]. A significant enhancement of the nonlinear response can be created by one of the numerous damages that might occur within concrete structures via quasi-static loading [30, 31], thermal stresses [8, 32, 33], carbonation [34], corrosion [35], and salt expansion [36]. Here, we show how different damage types affect the nonlinear observations derived from the scaling subtraction method, thus suggesting SSM to be a suitable method to monitor damage evolution in time. We also wish to highlight how nonlinear indicators defined using the SSM approach (particularly the slope b) allow to discriminate between different types of damage.

5.1 Load effects on discontinuities in concrete

One of the major effects that create damage in concrete is the application of mechanical loads in the presence of discontinuities. Indeed, discontinuity surfaces are very often the place from where damage may begin its progression. The effects could be the increase in crack density [37] and/or the crack openings [38, 39], depending on the nature of discontinuity such as existing cracks [40] or weak layers [41–43].

In [25], one specimen with an internal discontinuity surface was produced by piling up two concrete cubes (measuring 10 cm on each side). The two pieces were joined using a thin layer of cement paste. Concrete cubes were produced using a concrete mix with CEM II A-L 42.5 R cement, ordinary aggregates (max. size = 16 mm) and a water to cement ratio equal to 0.74, with no admixtures. Their age at the date of testing was approximately 6 months. The evaluation of the mechanical characteristics of the concrete was performed using mono-axial static compression tests that resulted in a compressive strength of 24 N/mm^2. The longitudinal wave speed in the cube was measured to be $V = 3850 m/s$ and the density of the cubes $\rho = 2330\ kg/m^3$. For this experiment, one emitter (T1) and three receivers (R2, R3, and R4) were used, as schematized in **Figure 3**. Two receivers (R2 and R3) performed direct transmission measurements; whereas, the third (R4) was used in indirect transmission mode. It is important to note that the direct path from the emitter to R2 and R4 crosses the discontinuity of the concrete, while the path to receiver R3 does not (See **Figure 3**). Only the results obtained from receiver R4 will be shown here for the sake of conciseness.

As explained above, the evolution of damage as a function of the applied load can be followed using different nonlinear indicators. However, in these experiments, it is important to note that the frequency analysis (using FFT for instance) was not efficient, since the nonlinear indicators related to the possible generated frequencies (higher order harmonics) were below the noise level. Therefore, the application of the traditional nonlinear elastic wave spectroscopy is not expected to be efficient in detecting the presence of nonlinearity during these experiments.

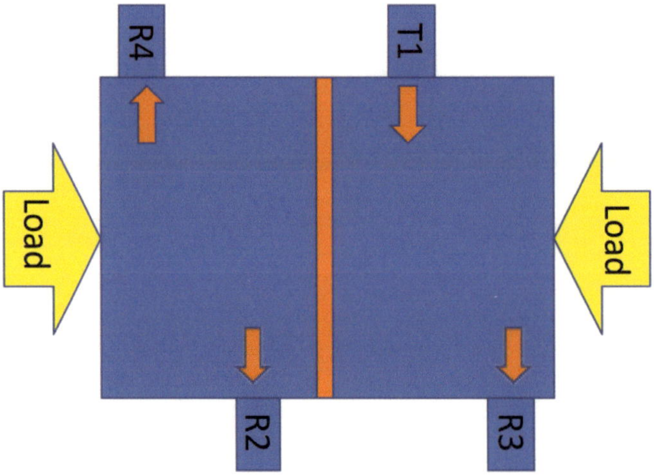

Figure 3.
Experimental set-up based on the use of four identical transducers, one transmitter (T1), and three receivers (R2, R_3, and R_4).

In general, when comparing different signals at different load levels, even if we observe a change in the signal shape, we should keep in mind that the change might not be due only to the damage of the interface, but also due to the other effects such as a change in the coupling quality or in the transducers arrangement especially when detaching and reattaching transducers. In that sense, it is important to note that the nonlinearity indication provided by the SSM represents an absolute measure since the reference signal is contained in the measurement itself.

Experimental results from [25] are shown in **Figure 4**. As anticipated above, the latter refers to signals recorded by the receiver R_4. The SSM indicator (θ) shows that the increase of load does not immediately cause a raise of nonlinearity, which starts to increase slightly only from 10 to 50 kN. However, we notice a clear increase in θ for higher loads, when the specimen is close to collapse, where fractures close to the bonding layer become evident around 90 kN. The log-log scale of θ on the same figure shows that θ changes linearly as a function of the amplitude. The slope of this logarithm evolution, which is around ~2 at low loading steps, becomes ~3 for higher load levels. Therefore, the evolution of nonlinearity as a function of damage progression can be appreciated, for a fixed value of the input energy, by analyzing the nonlinear indicator as a function of the applied load. This is done by extrapolating θ from the fitting function in order to obtain ($\theta(x_0)$) at each loading step.

The trend presented in **Figure 4** leads to the following conclusions. The low compressive loads create a rearrangement of the internal structure (e.g., pores closing) where the damage in the discontinuity is expected to be minor. When loads are increased (up to ~50% of the failure load), the nonlinearity increases slightly, and early damage (microdamage) is expected in the discontinuity. Finally, macrodamage is created at loads larger than ~50% of the failure load, where a clear change of the slope is observed.

5.2 Corrosion effect on elastic properties of reinforced concrete

Steel corrosion in reinforced concrete elements has negative effects on their load carrying capacity [44]. Furthermore, the expansion of the oxidation products creates cracking [45, 46] and deteriorates the bond between steel and concrete [47]. The overall weakening of the concrete structures by reinforcement corrosion is then expected [48].

DOI: http://dx.doi.org/10.5772/intechopen.82621

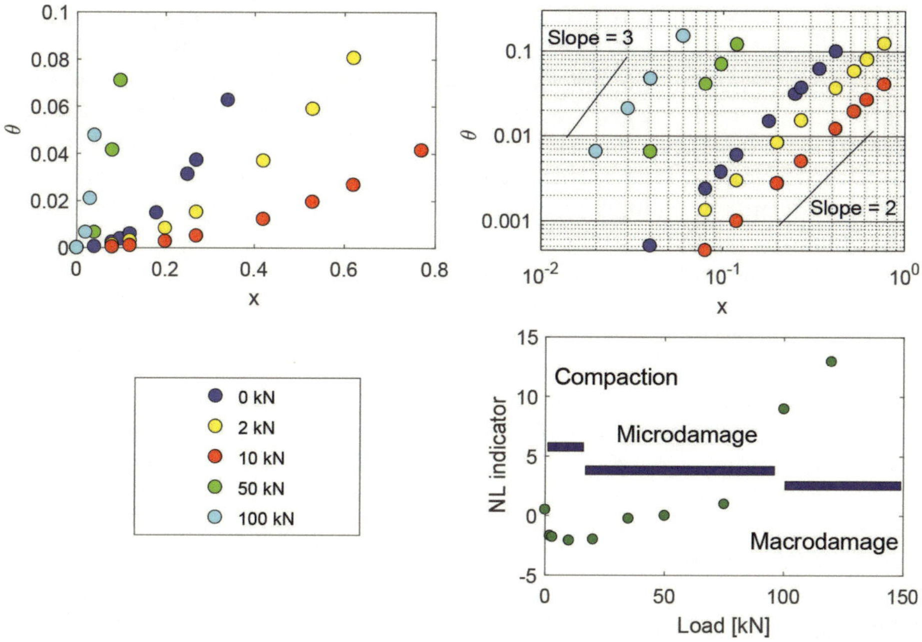

Figure 4.
Evolution of the SSM indicator θ as a function of energy of the output signal in the case of the low-quality layer discontinuity.

In [35], ultrasonic tests were performed on one reinforced concrete sample sizing (90 × 90 × 500 mm^3), which contained a steel bar (14 mm of diameter) located approximately in its center, making a reinforcement ratio corresponding to 0.19% (see **Figure 5**). Ultrasonic measurements consisted in exciting the sample using sinusoidal bursts at increasing amplitudes (10 cycles of 55.5 kHz frequency). Ultrasonic measurements were performed at different corrosion steps, as induced by an accelerated corrosion set-up, and data analysis was made on the basis of linear and nonlinear indicators. Linear indicators were the ultrasonic phase velocity (compression mode) and attenuation. Linear wave velocity was determined using the lowest excitation amplitude. Since transducers were removed before the beginning of each accelerated corrosion step, it was preliminarily verified that effects due to small differences in coupling were negligible compared to the ones induced by corrosion. Transducer positioning was made in such a way to be able to detect the direct transmission of the traveling ultrasonic wave. Furthermore, another transducer was put on the same side as the transmitter (at ~15 cm) in order to estimate the sensitivity of the proposed techniques when the access to the opposite face is not possible.

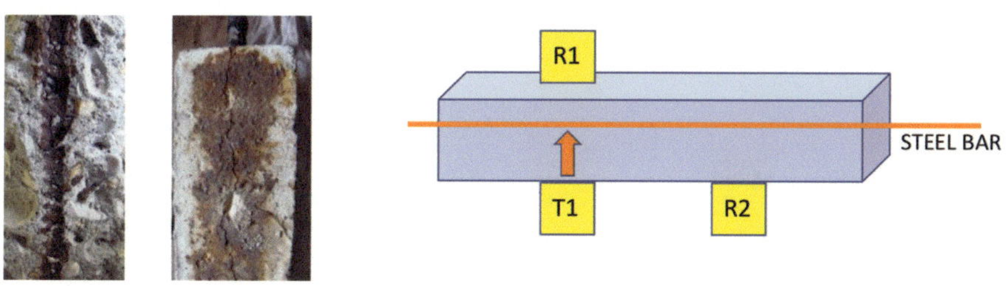

Figure 5.
Reinforced concrete samples affected by an important corrosion (left); ultrasonic transmitters and receivers placed on the concrete sample (right).

At the time when linear indicators, namely ultrasonic velocity and attenuation, manifested a weak sensitivity to corrosion (the velocity regularly but slightly decreases and the attenuation increases), the nonlinear indicator showed an important evolution. Indeed, the relative variations corresponding to velocity and attenuation as a function of the corrosion step were determined as 4 and 70% at most, respectively. However, larger effects were observed for the relative variation of the nonlinear indicator by changing up to 350%. Here, it should be pointed out that when comparing the first and the last corrosion steps (instead of considering as reference the intact sample), the relative change of the velocity and attenuation were 1.5 and 30%, respectively, while the nonlinear indicator was ~250%. This evolution shows the high sensitivity of nonlinear methods to the microstructure modifications due to corrosion. Finally, the nonlinear indicator remains sensitive to corrosion creation and evolution even when the access to the opposite side of the corroded concrete samples is not possible. Indeed, **Figure 6** shows that the nonlinear indicator extracted from data of receivers 1 and 2 was almost the same at the first two stages of corrosion. For the last two corrosion states, a difference can be noticed between both results, but the increase of sensitivity to corrosion is clearly visible for both sensors.

5.3 Thermally induced damage in concrete

Thermal stresses are a further cause of damage formation in concrete. To analyze the efficiency of the SSM in monitoring evolution of damage due to heating, mortar samples of size 25 × 25 × 100 mm^3 were prepared in [8]. Ordinary Portland cement (CEM I 42.5N) was used with aggregates from alluvial sand, with well-rounded particles and dry density of 2659 kg m^3. The amount of aggregates in the mortar was about 40% by volume, corresponding to a cement-to-aggregate ratio (c/a) by a mass fraction of 0.917. Water-to-cement ratio (w/c) was chosen as 0.3 by mass. Samples with larger and smaller grains were prepared.

Samples were heated in an oven at different temperature levels and ultrasonically monitored at each temperature level after having removed from oven and let cool down to room temperature. A linear analysis was applied measuring time of

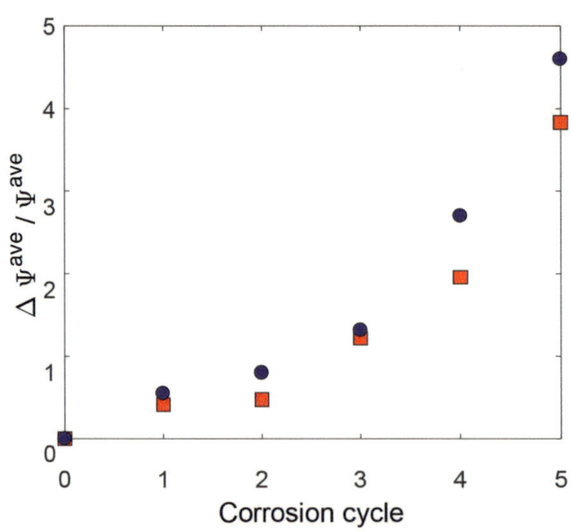

Figure 6.
Relative variations of the nonlinear indicator (with respect to the uncorroded initial state) as a function of the corrosion step. Squares and circles correspond to data recorded by receivers 1 and 2, respectively.

flight, but it proved to be only slightly sensitive to the increase of damage and without any difference in the behavior of samples with small and large aggregates sizes.

Thus, in order to appreciate the different behaviors expected for the different samples, the SSM technique was applied as discussed previously, with ultrasonic sensors attached on the bases of the sample prisms. Results are reported in **Figure 7**, where the nonlinear parameter is shown vs. the output amplitude. It is possible to observe that damage starts earlier in the sample with large aggregates, as expected. Macrocracks are more rapidly formed and at the largest thermal excitation, the increase in nonlinearity is noticeable.

Furthermore, it is also possible to observe a change in slope (nonlinear indicator b) when macrocracks start appearing (large aggregates case only). This is similar to what observed for the case of quasi-static loadings. Indeed, in both quasi-static and thermal cases, the situation is similar: microcracking (first) and coalescence into macrocracks (later) is due to the presence of localized mechanical stresses (due to load or local gradients of thermal expansion), without the formation of any reaction products, as in the cases of corrosion and salt expansion (see next subsection).

5.4 Degradation by expansive salts in masonry systems

The presence of soluble salts into capillary water is one of the major problems affecting masonry structures. Numerous works pointed out the potential noxiousness of water-transported salts (which might happen during repeated wet-dry cycles). Indeed, due to crystallization of some salts in the form of expansive compounds, progressive cracking and detachment phenomena happen especially at the interfaces between different material layers. In that sense, there is a clear need to develop an effective and reliable diagnosis of the onset of such damage phenomena in order to make corrective actions, in terms of repair and maintenance optimization, possible [49–53]. The close interaction between the ultrasonic wave and the material mechanical/elastic properties made ultrasonic methods widely used for

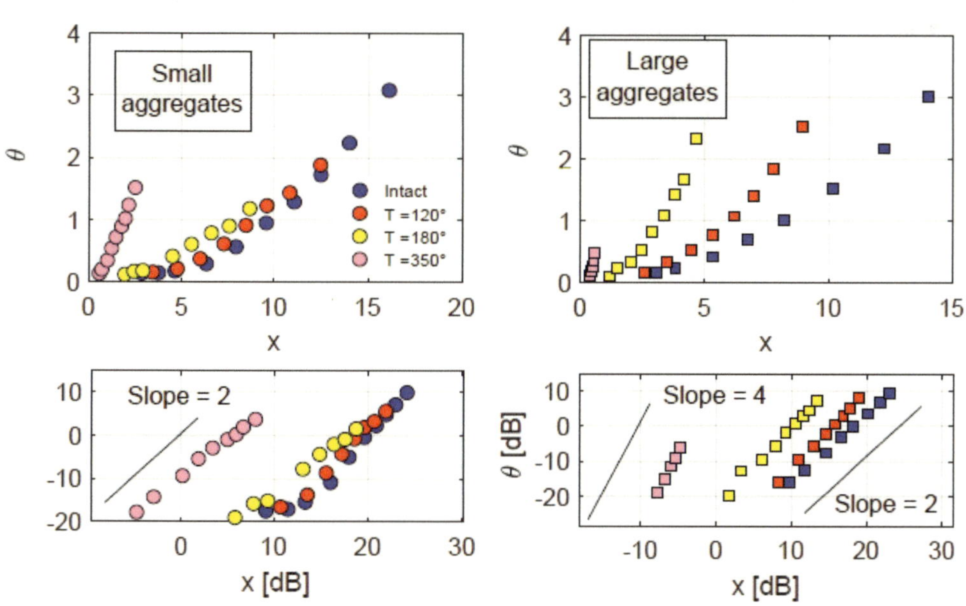

Figure 7.
Nonlinear parameter vs. output amplitude for samples with small and large aggregates at different levels of thermal damage.

damage characterization in civil engineering, in particular when dealing with cement-based materials. However, it has been remarked that ultrasonic pulse velocity, dispersion, and attenuation phenomena in concrete and multiphase materials in general are only considered in the linear regime at the time when without taking into account any possible dependence on the excitation amplitude [54–56]. Indeed, degradation of material mechanical properties can be more easily detected using nonlinear ultrasounds, which is significantly more sensitive at the very early damage stages than traditional linear ultrasonic techniques.

In [36], the sensitivity of the SSM was investigated, with respect to its application to the characterization of damage induced by the crystallization of expansive salts in masonry systems. Salt crystallization in coupled brick-mortar specimens was carried out through a damage protocol, which consisted of repeating cycles of imbibition in a salt solution followed by a drying phase. Specimens were manufactured by coupling a clay brick with two external layers of dehumidifying mortar using commercially available bricks in order to be representative of most common building products. A hand-molding procedure was used with natural drying and subsequent baking at the temperature of 1030°C. Irregular prismatic shape sizing $240 \times 115 \times 63\ mm^3$, lightly sanded on the outside surface, was then obtained. In analogy with typical restoration systems, a dehumidifying mortar was used for the external layers of the specimens in order to protect the underlying masonry from capillary rising dump, in the presence of soluble salts.

Recommendations from Standard UNI EN 12370 were used as a starting point for producing an accelerated decay protocol on the lab specimens. Then, the following damage protocol is applied.

- Immersion in a solution (14% of Na_2SO_4, by weight), for 2 h at $(20 \pm 0.5^{\circ}C)$.
- Heating at $85 \pm 5^{\circ}C$ during ~14 h.
- Heating at $105 \pm 5^{\circ}C$ during ~4 h.
- Cooling at room temperature during ~2 h.

Direct transmission of longitudinal ultrasonic waves was performed using two identical piezoelectric transducers with a central frequency of 55.5 kHz and 40 mm diameter (see scheme in **Figure 8**). Input bursts are composed of 10 sine cycles with a frequency of 55.5 kHz and fixed amplitude. Then, the excitation amplitude was progressively increased for successive bursts: for each damage step, the excitation amplitudes ranged between 1 and 20 V, before amplification. A time window of 200 ms was used for the acquisition of the received waveforms, and only first arrivals were considered in the analysis.

Figure 9 shows the evolution of the nonlinear effects caused by damage accumulation. Indeed, through the latter, it is possible to see the change in the rise of the nonlinear parameter as a function of the excitation energy when damage increases. Indeed, the power law evolution shows that the necessary amount of the excitation energy to trigger the nonlinear behavior decreases as a function of damage. Note that during the same damage steps, almost no mass variation occurred. This means that the brick-mortar adherence of the tested specimen ensured a very good resistance to damage growth during the first imbibition-drying cycles. On the other hand, the SSM analysis is in accordance with the mass variation measurements. Indeed, the semi-log plot of the same curve shows that the slope remains unchanged (approximately 2) during the first damage steps (partially cracked specimen), and the presence of first cracks (or microcracks) increases the sensitivity of the SSM

Figure 8.
Experimental set-up for ultrasonic characterization tests.

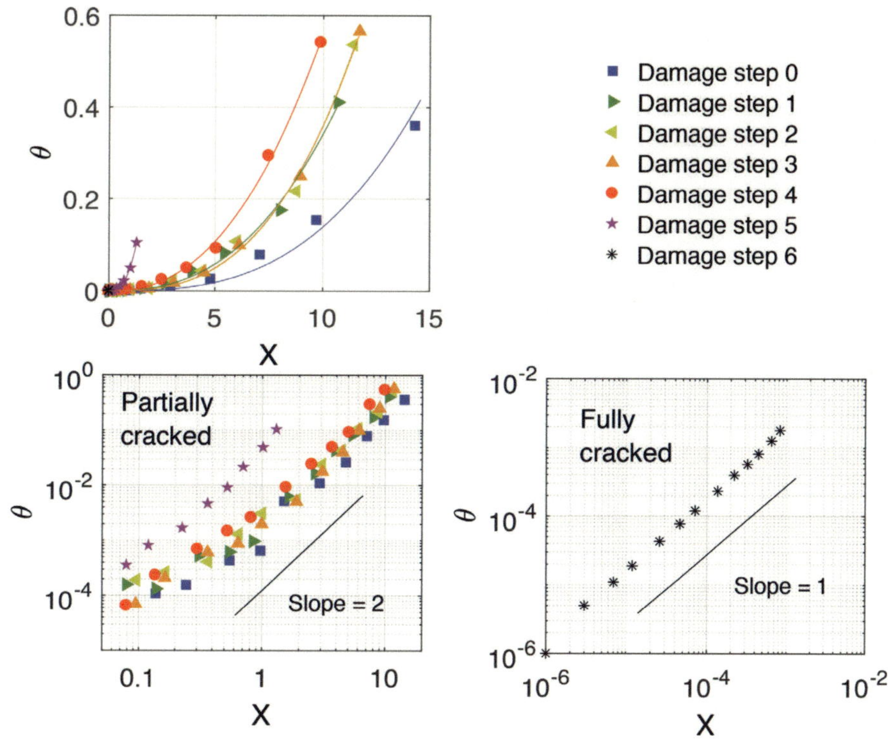

Figure 9.
Evolution of the nonlinear parameter vs. normalized energy at increasing degradation steps by expansive salts.

parameter without changing the power law behavior. At the end of the damage steps (fully cracked sample), a clear change of the slope can be noticed, since it goes from ~2 to ~1, again in accordance with the mass variation measurements.

6. Conclusion

In this chapter, we analyzed effects of different types of damage on the nonlinear properties of concrete using longitudinal direct transmission mode measurements. Our results suggest that, despite the damage origin (mechanical stress, thermal actions, corrosion, etc.), nonlinear ultrasonic measurements could be a useful tool for monitoring damage. In particular, the analysis performed in the time domain using the scaling subtraction method allows obtaining a sufficiently large signal-to-noise ratio, which is not always the case with standard analysis in the frequency domain. Furthermore, the high sensitivity to the appearance of degradation allows to provide precursory indications with approaching rupture. Further studies will be performed to exploit the possibility of using our approach to assess full-scale structures where, beyond the technical problems, we still need to perform a real strategy to choose the proper excitation frequency in order to overcome difficulties related to material attenuation.

Author details

Mourad Bentahar[1*], Charfeddine Mechri[1,2], Paola Antonaci[3], Antonio Gliozzi[4] and Marco Scalerandi[4]

1 LAUM, CNRS, Université du Maine, Le Mans, France

2 CTTM, Le Mans, France

3 Department of Structural, Geotechnical and Building Engineering, Politecnico di Torino, Torino, Italy

4 Department of Applied Science and Technology, Politecnico di Torino, Torino, Italy

*Address all correspondence to: mourad.bentahar@univ-lemans.fr

References

[1] Cantrell J, Yost W. Acoustic harmonic generation from fatigue induces dislocation dipoles. Philosophical Magazine. 1996;**69**: 315-326

[2] Moussatov A, Castagnéde B, Gusev V. Frequency up-conversion and frequency down-conversion of acoustic waves in damaged materials. Physics Letters A. 2002;**301**:281-290

[3] Zaitsev V, Nazarov V, Tournat V, Gusev V, Castagniéde B. Luxembourg-Gorky effect in a granular medium: Probing perturbations of the material state via cross-modulation of elastic waves. Europhysics Letters. 2005;**70**: 607-613

[4] Bentahar M, El Guerjouma R. Monitoring progressive damage in polymer-based composite using nonlinear dynamics and acoustic emission. The Journal of the Acoustical Society of America. 2009;**125**:39-44

[5] Muller M, Sutin A, Guyer R, Talmant M, Laugier P, Johnson P. Nonlinear resonant ultrasound spectroscopy (NRUS) applied to damage assessment in bone. The Journal of the Acoustical Society of America. 2005;**118**:3946-3952

[6] Van Den Abeele KE-A, Carmeliet J, Tencate J, Johnson P. Nonlinear elastic wave spectroscopy (NEWS) techniques to discern material damage. Part II: Single mode nonlinear resonance acoustic spectroscopy. Research in Nondestructive Evaluation. 2000;**12**:31-42

[7] Mechri C, Scalerandi M, Bentahar M. Enhancement of harmonics generation in hysteretic elastic media induced by conditioning. Communications in Nonlinear Science and Numerical Simulation. 2017;**45**:117-128

[8] Scalerandi M, Griffa M, Antonaci P, Wyrzykowski M, Lura P. Nonlinear elastic response of thermally damaged consolidated granular media. Journal of Applied Physics. 2013;**113**:154902

[9] Zinszner B, Johnson PA, Rasolofosaon PNJ. Influence of change in physical state on elastic nonlinear response in rock: Significance of effective pressure and water saturation. Journal of Geophysical Research. 1997; **102**:8105-8120

[10] Baccouche Y, Bentahar M, Mechri C, Novak A, El Guerjouma R. Nonlinear analysis of damaged metal-based composite plates using guided waves. Acta Acustica United with Acustica: Aims and Scope. 2017;**103**:967-977

[11] Bentahar M, El. Aqra H, El Guerjouma R, Griffa M, Scalerandi M. Hysteretic elasticity in damaged concrete: quantitative analysis of slow and fast dynamics. Physical Review B. 2006;**73**:014116

[12] Tencate J et al. Nonlinear and nonequilibrium dynamics in geomaterials. Physical Review Letters. 2004;**93**:065501

[13] Johnson PA, Sutin A. Slow dynamics and anomalous nonlinear fast dynamics in diverse solids. The Journal of the Acoustical Society of America. 2005;**117**: 124-130

[14] Aleshin V, Van Den Abeele K. Microcontact-based theory for acoustics in microdam-aged materials. Journal of the Mechanics and Physics of Solids. 2007;**55**:765-787

[15] Darling TW, Tencate JA, Sven V, et al. Localizing nonclassical nonlinearity in geological materials with neutron scattering experiments. In: Innovations in Nonlinear Acoustics: AIP Conference Proceedings. Vol. 838; 2006. pp. 12-19

[16] Johnson PA, Rasolofosaon PNJ. Manifestation of nonlinear elasticity in rock: Convincing evidence over large frequency and strain intervals from laboratory studies. Nonlinear Processes in Geophysics. 1996;**3**:77-88

[17] Scalerandi M, Delsanto PP. Modeling nonclassical nonlinearity, conditioning, and slow dynamics effects in mesoscopic elastic materials. Physical Review B. 2003;**68**:064107

[18] Shkerdin G, Glorieux C. Nonlinear modulation of lamb modes by clapping delamination. The Journal of the Acoustical Society of America. 2008; **124**:3397-3409

[19] Gusev V, Castagnède B, Moussatov M. Hysteresis in response of nonlinear bistable interface to continuously varying acoustic loading. Ultrasonics. 2003;**41**:643-654

[20] Baccouche Y, Bentahar M, Mechri C, El Guerjouma R. Hysteretic nonlinearity analysis in damaged composite plates using guided waves. Journal of Acoustical Society of America. 2013;**133**(4):EL256-EL261

[21] Bonner BP, Berge P, AracneâRuddle C, Boro C, Hardy E, Trombino C. Ultrasonic characterization of synthetic soils for application to near surface geophysics. In: Proceedings of the Symposium on the Application of Geophysics to Engineering and Environmental Problems; 1999. pp. 455-463

[22] Scalerandi M, Gliozzi AS, Bruno CLE, Antonaci P. Nonequilibrium and hysteresis in solids: Disentangling conditioning from nonlinear elasticity. Physical Review B. 2010;**81**:104114

[23] Bruno CLE et al. Analysis of elastic nonlinearity using the scaling subtraction method. Physical Review B. 2009;**79**:064108

[24] Shah AA, Ribakov Y. Non-destructive evaluation of concrete in damaged and undamaged states. Materials and Design. 2009;**30**: 3504-3511

[25] Antonaci P, Bruno CLE, Bocca PG, Scalerandi M, Gliozzi AS. Nonlinear ultrasonic evaluation of load effects on discontinuities in concrete. Cement and Concrete Research. 2010;**40**:340-346

[26] Hilloulin B et al. Small crack detection in cementitious materials using nonlinear coda wave modulation. NDT and E International. 2014;**68**: 98-104

[27] Quang AV et al. Concrete cover characterization using dynamic acousto-elastic testing and Rayleigh waves. Construction and Building Materials. 2016;**114**:87-97

[28] Aggelis D, Momoki S, Shiotani T. Experimental study of nonlinear wave parameters in mortar. Construction and Building Materials. 2013;**47**:1409-1413

[29] Garnier V et al. Acoustic techniques for concrete evaluation: Improvements, comparisons and consistency. Construction and Building Materials. 2013;**43**:598-613

[30] van den Abeele K, De Visscher J. Damage assessment in reinforced concrete using spectral and temporal nonlinear vibration techniques. Cement and Concrete Research. 2000;**30**: 1453-1464

[31] Antonaci P, Bruno CLE, Gliozzi AS, Scalerandi M. Monitoring evolution of compressive damage in concrete with linear and nonlinear ultrasonic methods. Cement and Concrete Research. 2010; **40**:1106-1113

[32] Payan C et al. Quantitative linear and nonlinear resonance inspection techniques and analysis for material characterization: Application to

concrete thermal damage. The Journal of the Acoustical Society of America. 2014;**136**:537-546

[33] Park S-J, Yim HJ, Kwak H-G. Effects of post-fire curing conditions on the restoration of material properties of fire-damaged concrete. Construction and Building Materials. 2015;**99**:90-98

[34] Bouchaala F, Payan C, Garnier V, Balayssac JP. Carbonation assessment in concrete by nonlinear ultrasound. Cement and Concrete Research. 2011; **41**:557-559

[35] Antonaci P, Bruno CLE, Scalerandi M, Tondolo F. Effects of corrosion on linear and nonlinear elastic properties of reinforced concrete. Cement and Concrete Research. 2013;**51**:96-103

[36] Antonaci P, Formia A, Gliozzi AS, Scalerandi M, Tulliani JM. Diagnostic application of nonlinear ultrasonics to characterise degradation by expansive salts in masonry systems. NDT and E International. 2013;**55**:57-63

[37] Paliwal B, Ramesh KT. An interacting micro-crack damage model for failure of brittle materials under compression. Journal of the Mechanics and Physics of Solids. 2008;**56**:896-923

[38] Galvez JC, Cervenka J, Cendonc DA, Saouma V. A discrete crack approach to normal/shear cracking of concrete. Cement and Concrete Research. 2002;**32**:1567-1585

[39] Zhang J, Li VC. Simulation of crack propagation in fiber-reinforced concrete by fracture mechanics. Cement and Concrete Research. 2004;**34**:333-339

[40] Barpi F, Valente S. Modeling water penetration at dam-foundation joint. Engineering Fracture Mechanics. 2008; **75**:629-642

[41] Djazmati B, Pincheira JA. Shear stiffness and strength of horizontal construction joints. ACI Structural Journal. 2004;**101**:484-493

[42] Sharif A, Rahman MK, Al-Gahtani AS, Hameeduddin M. Behaviour of patch repair of axially loaded reinforced concrete beams. Cement and Concrete Composites. 2006;**28**:734-741

[43] Mangat PS, O'Flaherty FJ. Influence of elastic modulus on stress redistribution and cracking in repair patches. Cement and Concrete Research. 2000;**30**:125-136

[44] Cairns J, Zhao Z. Behavior of concrete beams with exposed reinforcement. Journal of Structural Engineering. 1999;**99**:141-154

[45] Seok JB, Hwan OB. Effects of non-uniform corrosion on the cracking and service life of reinforced concrete structures. Cement and Concrete Research. 2010;**40**:1441-1450

[46] Care S et al. Times to cracking in reinforced mortar beams subjected to accelerated corrosion tests. Materials and Structures. 2010;**43**:107-124

[47] Fang C et al. Corrosion influence on bond in reinforced concrete. Cement and Concrete Research. 2004;**34**: 2159-2167

[48] Masayasu O, Uddin FAKM. Mechanisms of corrosion-induced cracks in concrete at meso- and macro-scales. Journal of Advanced Concrete Technology. 2008;**6**:419-429

[49] Rirsch E, Zhang Z. Rising damp in masonry walls and the importance of mortar properties. Construction and Building Materials. 2010;**24**:1815-1820

[50] Lubelli B, Van Hees RPJ, Groot CJWP. The role of sea salts in the occurrence of different damage mechanisms and decay patterns on brick masonry. Construction and Building Materials. 2004;**18**:119-124

[51] Brocken H, Nijland TJ. White efflorescence on brick masonry and concrete masonry blocks, with special emphasis on sulfate efflorescence on concrete blocks. Construction and Building Materials. 2004;**18**:315-323

[52] Fassina V et al. Evaluation of compatibility and durability of a hydraulic lime-based plaster applied on brick wall masonry of historical buildings affected by rising damp phenomena. Journal of Cultural Heritage. 2002;**3**:45-51

[53] Abu-Zei N et al. Non-invasive characterization of ancient foundations in Venice using the electrical resistivity imaging technique. NDT and E International. 2006;**39**:67-75

[54] Komlos K et al. Ultrasonic pulse velocity test of concrete properties as specified in various standards. Cement and Concrete Composites. 1996;**18**: 357-364

[55] Kim YH, Lee S, Kim HC. Attenuation and dispersion of elastic waves in multi-phase materials. Journal of Physics D. 1991;**24**:1722-1728

[56] Landis E, Shah S. Frequency-dependent stress wave attenuation in cement-based materials. Journal of Engineering Mechanics. 1995;**121**: 737-743

Chapter 8

Resonance Compression of Acoustic Beams in Crystals

Vladimir I. Alshits, Dmitrii A. Bessonov and Vasilii N. Lyubimov

Abstract

The resonant excitation of an intense elastic wave through nonspecular reflection of a special pump wave in a crystal is described. Geometric criteria are found under which mode conversion, when the incident and reflected beams belong to different acoustic branches, coexists with total internal reflection of an acoustic beam. In this case, the entire energy of an incident pump wave is spent on the excitation of a narrow intense reflected beam close in structure to an eigenmode. A consistent choice of orientations of the sagittal plane and crystal surface that excludes the reflection of a parasitic wave of leakage is found. The resonance parameters have been found for a medium with an arbitrary anisotropy. General relations are concretized for monoclinic, orthorhombic, trigonal, tetragonal, cubic, and hexagonal systems. Estimates and illustrations are given for a series of such crystals. The intensity of the reflected beam increases with its narrowing, but its diffraction divergence also increases with this narrowing. Nevertheless, the intensity of the beam can be increased by a factor of 5–10 at sufficiently high frequencies while keeping its divergence at an acceptable level. Amplification by two orders of magnitude can be achieved by compressing the beam in two dimensions through its double reflection.

Keywords: crystals, elastic waves, acoustic beams, total internal reflection, mode conversion, efficiency of transformation, diffraction divergence

1. Introduction

Modern crystal acoustics is an important base for numerous instruments and devices using concentrated ultra- and hypersonic beams, delay lines, surface and bulk waves, etc. Many of acoustic effects in crystals arise exclusively due to their anisotropy. In particular, piezoelectricity exists only in crystals and is widely used in acoustic devices [1, 2]. Another spectacular example of a nontrivial role of anisotropy is phonon focusing [3], the concentration of energy in a crystal along special directions for which the acoustic beam in Poynting vectors is much narrower than that in wave vectors. In this chapter, we will consider another principle of energy concentration in acoustic waves that is also entirely related to crystal anisotropy. Intense ultrasonic beams are widely used in engineering, medicine, scientific instrument making, etc. [4]. The reflection and refraction of such beams at the interfaces between layered isotropic structures are commonly used for their transformation. Crystals open up new opportunities for beam transformation.

A method for producing intense beams in crystals based on the features of their elastic anisotropy was proposed in our paper [5]. "Compression" of an acoustic beam is achieved by choosing the geometry of the beam incidence on a surface close to the angle of total internal reflection, when one of the reflected beams ($r2$) propagates at a small angle to the surface (**Figure 1a**) and, its width can be made arbitrarily small. However, we need the compression not of the beam width, but of its energy density. For instance, in isotropic medium, a compression of such beam is accompanied by a decreasing amount of energy entering it, without any growth of its intensity. The same occurs in the crystal without a special choice of reflection geometry. And still in anisotropic media, there are specific orientations which admit the beam intensification.

As is shown in [5], this happens when the wave field of the beam $r2$ is close to the eigenmode—an exceptional bulk wave (EBW) satisfying the free surface boundary condition [6, 7]. The perturbation of the selected EBW propagation geometry transforms this one-partial solution to the resonance reflection component. To obtain such a special resonance reflection near the eigenmode, the EBW should exist on the middle sheet of the slowness surface, while the incident "pump" wave should belong to the external sheet (**Figure 1b**). The proper cuts can be found almost in any crystal. However, one should bear in mind that in this case, apart from the reflected wave $r2$ excited from the middle sheet of the slowness surface, another reflected wave $r1$ belonging to the external sheet inevitably exists. Energy losses related to this parasitic wave can be minimized by choosing crystals or geometries with parameters corresponding to the closeness of reflection to mode conversion, when such a parasitic wave does not appear.

In [5, 8], we considered perturbed geometries, where the surface of the crystal remained unchanged and the plane of reflection (sagittal plane) was rotated by a small angle φ about the normal $\mathbf{n}$ to the surface (**Figure 1a**). In this case, an increase in the intensity of excited beam was controlled by the angle φ, whereas the energy loss to the parasitic beam was completely determined by the relation between the moduli of elasticity and could be reduced only by an appropriate choice of the crystal.

In [9], we analyzed another variant of the theory where a similar resonance in a hexagonal crystal was governed by the angle of rotation of the surface about the direction of propagation of the unperturbed exceptional bulk wave. In this case, conversion also occurs only under a certain relation between moduli of elasticity.

In [10, 11], the more general analysis was accomplished which allowed us to demonstrate that the mode conversion of resonance near total internal reflection (i.e., the scheme in **Figure 1a** without the parasitic beam $r1$) can be implemented in almost any acoustic crystal by a consistent variation of orientations of both the boundary and the sagittal planes.

In this chapter, we shall summarize the results of mentioned and some other studies of ours and present the combined theoretical consideration of the problem with both analytical approximate calculations and numerical exact computations

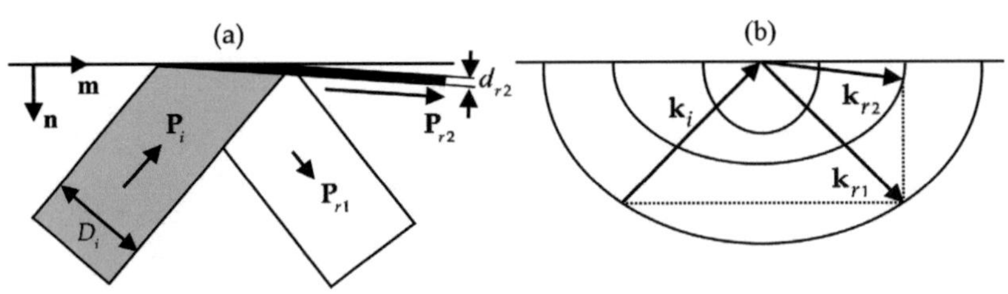

Figure 1.
Scheme of excitation of a narrow beam near total internal reflection in the $\mathbf{r}$ *space (a) and* $\mathbf{k}$ *space (b) with* $\mathbf{P}_i$, $\mathbf{k}_i$; $\mathbf{P}_{r1}$, $\mathbf{k}_{r1}$; *and* $\mathbf{P}_{r2}$, $\mathbf{k}_{r2}$ *to be mean Poynting and wave vectors of the incident, parasitic, and excited beams, respectively.*

checking the established relations between the basic parameters determining the unusual resonance phenomenon with features quite promising for applications.

2. Formulation of the problem and basic relations

Consider a semi-infinite elastic medium of unrestricted anisotropy with a free boundary. It will be characterized by the tensor of moduli of elasticity c_{ijkl} and the density ρ. The sagittal plane is specified by two unit vectors: the propagation direction **m** along the surface and the normal **n** to the surface. Reflection shown schematically in **Figure 1b** is the superposition of four partial waves: the incident ($\alpha = 4 = i$) and reflected ($\alpha = 1 = r1$) waves from the outer sheet of the slowness surface, the reflected wave ($\alpha = 2 = r2$) from the middle sheet, and the localized wave ($\alpha = 3 = l$) from the internal sheet:

$$\begin{pmatrix} \mathbf{u}(\mathbf{r},t) \\ (i/k)\hat{\sigma}(\mathbf{r},t) \end{pmatrix} = \sum_{\alpha=1}^{4} C_\alpha \begin{pmatrix} \mathbf{A}_\alpha \\ \mathbf{L}_\alpha \end{pmatrix} \exp\{i(\mathbf{k}_\alpha \cdot \mathbf{r} - \omega t)\}. \quad (1)$$

Here, **u** is the wave field displacement vector, $\hat{\sigma}$ is the stress tensor, C_α are the partial waves amplitudes, ω is the frequency, and $\mathbf{k}_\alpha = k(\mathbf{m} + p_\alpha \mathbf{n})$ are the wave vectors of partial components with a common projection k onto the direction of propagation **m** (**Figure 1b**). The k value determines the tracing speed $v = \omega/k$ of stationary motion of the wave field (1) along the boundary. Vectors $\mathbf{A}_\alpha$ and $\mathbf{L}_\alpha$ characterize the partial field polarizations. Being not independent (as well as **u** and $\hat{\sigma}$), these vectors are normalized by the condition: $\mathbf{A}_\alpha^2 = 1$.

In terms of Eq. (1), the boundary condition of free surface, $\sigma_{ij} n_j\big|_{y=0} = 0$, takes the form:

$$C_{r1}\mathbf{L}_1 + C_{r2}\mathbf{L}_2 + C_l\mathbf{L}_3 + C_i\mathbf{L}_4 = 0. \quad (2)$$

The unknown vectors $\mathbf{A}_\alpha$ and $\mathbf{L}_\alpha$ are found from the so-called Stroh's formalism based on the fact that the combined six-vectors $\xi_\alpha = \{\mathbf{A}_\alpha, \mathbf{L}_\alpha\}^T$ (the superscript T means transposition) together with parameters p_α ($\alpha = 1,\ldots,6$) are eigenvectors and eigenvalues of the 6×6 Stroh matrix $\hat{N}$ [12],

$$\hat{N} = -\begin{pmatrix} (nn)^{-1}(nm) & (nn)^{-1} \\ (mn)(nn)^{-1}(nm) - (mm) - \rho v^2 \hat{I} & (mn)(nn)^{-1} \end{pmatrix}. \quad (3)$$

Here $\hat{I}$ is the unit 3×3 matrix and the matrices (ab) are defined by the convolutions $(ab)_{jk} = a_i c_{ijkl} b_l$ of the moduli tensor c_{ijkl} with the vectors **a** and **b**. The six eigenvectors ξ_α are complete and orthogonal to each other everywhere apart from points of degeneracy. The orthogonality property may be expressed in the form:

$$\mathbf{A}_\alpha \cdot \mathbf{L}_\beta + \mathbf{A}_\beta \cdot \mathbf{L}_\alpha = 0, \quad \alpha \neq \beta. \quad (4)$$

Depending on v, the vectors ξ_α and the parameters p_α may be real or form complex conjugated pairs. The reflection considered in this chapter (**Figure 1b**) belongs to the second supersonic region of the slowness surface. In the above terms, here the wave superposition formally may include four bulk partial waves with real parameters p_α, two incident and two reflected, from the external and middle sheets. In addition, at our disposal, there are two inhomogeneous partial waves with complex conjugated parameters p_α, one localized and the other nonphysical (increasing into the depth of the medium), related to the internal sheet. The second incident wave and the nonphysical inhomogeneous component were naturally excluded ($C_5 = C_6 = 0$) from the sums in Eqs. (1) and (2).

The amplitude C_i of the incident wave is assumed to be known, while the remaining amplitudes may be expressed in terms of C_i through scalar multiplication of Eq. (2) by the vector products $\mathbf{L}_2 \times \mathbf{L}_3$, $\mathbf{L}_1 \times \mathbf{L}_3$, or $\mathbf{L}_1 \times \mathbf{L}_2$. As a result, we arrive at the following reflection coefficients in the form of the ratios of mixed products:

$$R_1 = \frac{C_{r1}}{C_i} = -\frac{[\mathbf{L}_4\mathbf{L}_2\mathbf{L}_3]}{[\mathbf{L}_1\mathbf{L}_2\mathbf{L}_3]}, \qquad R_2 = \frac{C_{r2}}{C_i} = \frac{[\mathbf{L}_4\mathbf{L}_1\mathbf{L}_3]}{[\mathbf{L}_1\mathbf{L}_2\mathbf{L}_3]}. \tag{5}$$

3. Singular reflection geometry in the vicinity of EBW

The deduced equations (5) represent just exact relations which describe the discussed resonance reflection only in some definite narrow region of orientations of surface and sagittal plane. Let us demonstrate that such singular region really arises close to the geometry of EBW propagation related to the pair $\{\mathbf{n}_0, \mathbf{m}_0\}$. Indeed, wave superposition (1) in this exceptional geometry is decomposed into two independent solutions—EBW and three-partial reflection—satisfying the boundary conditions, which are fragments of Eq. (2):

$$\mathbf{L}_{02} = 0, \tag{6}$$

$$C_{r1}\mathbf{L}_{01} + C_l\mathbf{L}_{03} + C_i\mathbf{L}_{04} = 0, \tag{7}$$

where the subscript 0 indicates the initial unperturbed configuration. Of course, near this geometry, the vector $\mathbf{L}_2$ should be small and the other vectors $\mathbf{L}_1$, $\mathbf{L}_3$, and $\mathbf{L}_4$ nearly coplanar. As a result, both the numerator and denominator of the expression for R_2 in (5) should independently approach zero; i.e., R_2 is singular. This singularity is responsible for the resonance character of reflection and, therefore, for the discussed effect. The coefficient R_1 in (5) is regular because the small vector $\mathbf{L}_2$ appears in both the denominator and the numerator of the expression for this coefficient.

Let us introduce practically important characteristics of the investigated resonance, specifically, the gain of excited wave intensity (K_2) and energy loss (K_1) in a parasitic wave:

$$K_2 = \frac{P_2}{P_4} = |R_2|^2\frac{s_2}{s_4}, \qquad K_1 = \frac{P_1}{P_4} = |R_1|^2\frac{s_1}{s_4}. \tag{8}$$

Here, P_α are the Poynting vector lengths, which are products of the energy density in the corresponding partial wave ($\propto C_\alpha^2$) and its ray speed s_α. Another useful characteristic of the resonance is the excitation efficiency $\eta = 1 - K_1$, equal to the fraction of energy transferred from the incident to excited wave.

Thus, a small perturbation of the crystal orientation in the vicinity of the geometry of the EBW propagation may provide a resonance intensification of the reflected wave $r2$. The control parameters of resonance are the optimized characteristics of geometry of reflection $\{\mathbf{n}, \mathbf{m}\}$ (**Figure 1a**), i.e., the orientations of the surface and sagittal plane, as well as the angle of incidence α of the pump wave i related to these characteristics. The perturbation $\{\mathbf{n}_0, \mathbf{m}_0\} \to \{\mathbf{n}, \mathbf{m}\}$ is shown in **Figure 2a**. The new plane boundary P of the crystal is specified by the unit normal $\mathbf{n} = \mathbf{n}(\psi, \chi)$ rotated by a small angle $\psi = \angle(\mathbf{n}, \mathbf{n}_0)$ with respect to the normal $\mathbf{n}_0$ to the initial boundary P_0 along which the EBW can propagate. The axis of rotation

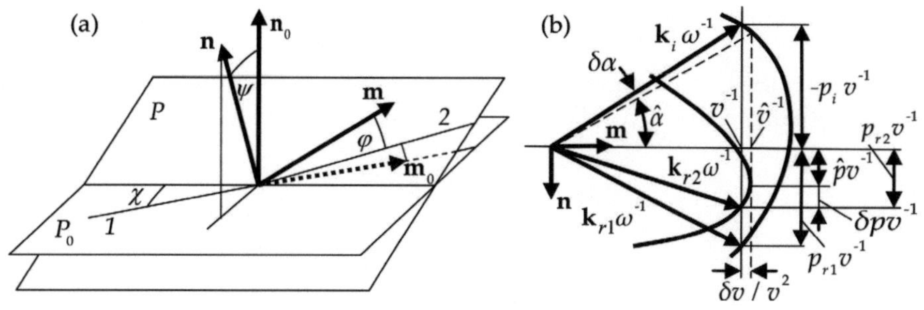

Figure 2.
(a) Perturbations of the crystal surface $P_0 \to P$*, its normal orientation* $\mathbf{n}_0 \to \mathbf{n}$ *and the propagation direction* $\mathbf{m}_0 \to \mathbf{m}$*; (b) the scheme of the resonance reflection (only outer and middle sheets of the slowness surface are shown).*

(the intersection line between P and P_0 planes) is specified by the angle χ, counting from reference line $1||\mathbf{m}_0$ and lying in the range of $[0, \pi]$. The direction $\mathbf{m}$ in the P plane is specified by the angle φ measured from line 2 along the projection of the unperturbed vector $\mathbf{m}_0$ on the boundary P.

Below, to compact equations, we will also use the alternative notation for the introduced small angles:

$$\phi_1 = \varphi, \qquad \phi_2 = \psi. \tag{9}$$

In fact, as we shall see, the squares of these angles (in units of radians) rather than the angles themselves are small parameters in the theory developed below.

As is seen in **Figure 2b**, where the scheme of resonance reflection is shown, the angle α of incidence must be chosen near the threshold angle of total internal reflection $\hat{\alpha}$: $\alpha = \hat{\alpha} + \delta\alpha$. This configuration is characterized by the tracing speed $v = \omega/k$ of the wave field close to the limiting velocity $\hat{v}$. The adjusting angle $\delta\alpha$ controlling reflection resonance is directly related to the difference $\delta v = v - \hat{v} \approx \delta p^2/2\kappa$ [8]:

$$\delta\alpha = \frac{\delta v}{\hat{v}} A = \frac{A}{2\kappa\hat{v}} \delta p^2, \qquad A = \frac{\hat{v}}{1 + p_i^2} \left(\frac{\partial p_i}{\partial v} \right)_{v = \hat{v}}, \tag{10}$$

where κ is the radius of curvature of the cross section of the middle sheet of the slowness surface by the sagittal plane at the limiting point corresponding to $v = \hat{v}$.

4. Analytical description of resonance characteristics

In this section, we shall obtain approximate expressions for the coefficients K_1 and K_2 defined by Eq. (8) in the general case of arbitrary anisotropy. Reflectivities R_1 and R_2 are expressed in (5) in terms of vectors $\mathbf{L}_\alpha$, which will be considered as functions of the parameters δp, φ, ψ, and χ (see **Figure 2**). The last (not small) angle χ, which determines the axis for $\mathbf{n}$ rotation, is considered fixed at this stage. Its influence on the effect will be studied numerically. Expressions (5) can be expanded in the other small parameters.

Given (7), the numerator in Eq. (5) for R_2 can be approximately represented as:

$$[\mathbf{L}_4 \mathbf{L}_1 \mathbf{L}_3] \approx \phi_i \left(\frac{\partial}{\partial \phi_i} [\mathbf{L}_4 \mathbf{L}_1 \mathbf{L}_3] \right)_0, \tag{11}$$

where the subscript 0 means that after differentiation, one should put $\phi = \psi = 0$, $\delta p = 0$.

To calculate the denominator in the expressions for reflection coefficients R_1 and R_2 (5), let us expand the two cofactors separately:

$$\mathbf{L}_3 \times \mathbf{L}_1 \approx \mathbf{L}_{03} \times \mathbf{L}_{01} + \phi_i \left(\frac{\partial}{\partial \phi_i} [\mathbf{L}_3 \times \mathbf{L}_1] \right)_0, \tag{12}$$

$$\mathbf{L}_2 \approx \phi_i \left(\frac{\partial \mathbf{L}_2}{\partial \phi_i} \right)_0 + \frac{1}{2} \phi_i \phi_j \left(\frac{\partial^2 \mathbf{L}_2}{\partial \phi_i \partial \phi_j} \right)_0 + \delta p \left(\frac{\partial \mathbf{L}_2}{\partial p_2} \right)_0. \tag{13}$$

In writing (13), we used the fact that $\mathbf{L}_{02} = 0$ (6). Substituting Eqs. (12) and (13) into the denominators of ratios (5), it is easy to verify that the term linear in φ and ψ drops out of the result. This term is proportional to the mixed product, which is zero,

$$\left[\mathbf{L}_{03} \mathbf{L}_{01} \left(\frac{\partial \mathbf{L}_2}{\partial \phi_i} \right)_0 \right] = 0, \tag{14}$$

because, as we will see, it is composed of coplanar vectors. Let us prove that they are all perpendicular to the same real vector $\mathbf{A}_{02}$. From orthogonality condition (4), one has

$$\mathbf{A}_{02} \cdot \mathbf{L}_{0\alpha} + \mathbf{L}_{02} \cdot \mathbf{A}_{0\alpha} = \mathbf{A}_{02} \cdot \mathbf{L}_{0\alpha} = 0. \tag{15}$$

for all $\alpha \neq 2$ where the fact that $\mathbf{L}_{02} = 0$ was again used.

In order to prove that the derivative $(\partial \mathbf{L}_2 / \partial \phi_i)_0$ is also orthogonal to $\mathbf{A}_{02}$, one can use identity $\mathbf{A}_2(\hat{v}) \cdot \mathbf{L}_2(\hat{v}) = 0$ valid for any transonic states of arbitrary geometry $\{\mathbf{m}, \mathbf{n}\}$ [13]. Let us differentiate this identity and then set $\phi = \psi = 0$:

$$\left(\mathbf{A}_2 \cdot \frac{\partial \mathbf{L}_2}{\partial \phi_i} + \mathbf{L}_2 \cdot \frac{\partial \mathbf{A}_2}{\partial \phi_i} \right)_0 = \mathbf{A}_{02} \cdot \left(\frac{\partial \mathbf{L}_2}{\partial \phi_i} \right)_0 = 0. \tag{16}$$

Thus, Eqs. (15) and (16) prove vanishing in (14), which means that indeed the denominator of both reflection coefficients (5) does not contain terms linear in φ and ψ. The numerator of R_1 is found from the same relations (12) and (13) after replacing in them indices $1 \to 4$, and the numerator of R_2 is given by Eq. (11). After some straightforward calculations, one obtains

$$R_1 = -\frac{F^{(4)} \delta p + G_{ij}^{(4)} \phi_i \phi_j}{F^{(1)} \delta p + G_{ij}^{(1)} \phi_i \phi_j}, \qquad R_2 = \frac{H_i \phi_i}{F^{(1)} \delta p + G_{ij}^{(1)} \phi_i \phi_j}, \tag{17}$$

where the repeated subscripts imply summation, and the new notations are introduced:

$$F^{(\alpha)} = \left[\mathbf{L}_\alpha \frac{\partial \mathbf{L}_2}{\partial p_2} \mathbf{L}_3 \right]_0, \qquad H_i = \left(\frac{\partial}{\partial \phi_i} [\mathbf{L}_4 \mathbf{L}_1 \mathbf{L}_3] \right)_0, \tag{18}$$

$$G_{ij}^{(\alpha)} = \left[\frac{\partial \mathbf{L}_2}{\partial \phi_i} \cdot \frac{\partial (\mathbf{L}_3 \times \mathbf{L}_\alpha)}{\partial \phi_j} + \frac{1}{2} \left(\mathbf{L}_\alpha \frac{\partial^2 \mathbf{L}_2}{\partial \phi_i \partial \phi_j} \mathbf{L}_3 \right) \right]_0. \tag{19}$$

For further compactness of expressions, let us also denote:

$$G_{ij}^{(\alpha)}/F^{(\alpha)} = \lambda_{ij}^{(\alpha)} \equiv \lambda'^{(\alpha)}_{ij} + i\lambda''^{(\alpha)}_{ij}, \qquad H_i/F^{(1)} = \mu_i. \tag{20}$$

The substitution of Eq. (17) in terms of (20) into Eq. (8) gives

$$K_1 = \frac{s_1}{s_4}\left|\frac{F^{(4)}}{F^{(1)}}\right|^2 \frac{\left(\delta p + \lambda'^{(4)}_{ij}\phi_i\phi_j\right)^2 + \left(\lambda''^{(4)}_{ij}\phi_i\phi_j\right)^2}{\left(\delta p + \lambda'^{(1)}_{ij}\phi_i\phi_j\right)^2 + \left(\lambda''^{(1)}_{ij}\phi_i\phi_j\right)^2}, \tag{21}$$

$$K_2 = \frac{(\mu_i\phi_i)^2 s_2/s_4}{\left(\delta p + \lambda'^{(1)}_{ij}\phi_i\phi_j\right)^2 + \left(\lambda''^{(1)}_{ij}\phi_i\phi_j\right)^2}. \tag{22}$$

As could be expected, loss coefficient K_1 (21) is regular in the control parameters φ, ψ, and δp, whereas gain K_2 (22) is singularly dependent on them: at $\phi_i\phi_j \ll \delta p \ll 1$, K_2 tends to zero, and at $\delta p \ll \phi_i\phi_j \ll 1$, it diverges.

5. Optimization of control parameters of reflection

Until now, we were free with a choice of geometry of the considered resonance reflection. It looks natural to choose the parameters φ, ψ, and δp so that the loss coefficient K_1 would be as small as possible, i.e., the efficiency η would be close to 100%. In terms of **Figure 1**, this means an exclusion of the parasitic reflected beam $r1$ which is equivalent to a realization of the mode conversion. Formula (21) allows reducing the criterion of conversion $K_1 = 0$ to the system of equations:

$$\lambda''^{(4)}_{ij}\phi_i\phi_j = 0, \quad \delta p = -\lambda'^{(4)}_{ij}\phi_i\phi_j, \tag{23}$$

where the first equation determines the relation between the angles of rotation of the sagittal plane ($\phi_1 \equiv \varphi$) and the normal to the surface ($\phi_2 \equiv \psi$) at a fixed position (χ) of the axis of rotation of the vector **n** (see **Figure 2a**). The second equation in (23) at the found relation between φ and ψ specifies the dependence $\delta p(\varphi)$ and, by (10), the incidence angle $\delta\alpha(\varphi)$.

The first requirement in (23) is reduced to a quadratic equation with respect to the ratio ψ/φ. The existence of real roots of this equation (and, therefore, mode conversion) is generally not guaranteed. However, numerical calculations for a number of crystals of different symmetry systems did not give us examples of the absence of such roots. Furthermore, as is shown in the next section, for hexagonal crystals, this equation always has real roots for the case $c_{44} > c_{66}$. Thus, in many crystals, the consistent variation of orientations of the surface and sagittal plane really can provide the mode conversion near the total internal reflection, i.e., the effect which is under consideration.

The general conditions of mode conversion (23) can be represented in the compact form:

$$\psi = \gamma_\pm \varphi, \quad \delta p = \lambda_\pm \varphi^2. \tag{24}$$

These conditions with real roots $\gamma_\pm$ specify two variants of the orientations of the surface, sagittal plane, and angle of incidence (for each angle χ, see **Figure 2a**) that ensure the energy concentrating in the reflected beam $r2$.

As was expected, the gain K_2 given by Eq. (22) should obviously be large because it is inversely proportional to the square of the small parameter:

$$K_2^{\rm con} \propto \varphi^{-2} \propto \psi^{-2}. \tag{25}$$

But the unlimited increase in (25) with a decrease in the angle φ should not mislead us. Indeed, an increase in the amplitude (25) of the resonance peak (22) is accompanied by its narrowing. However, when this width in angles of incidence $\delta\alpha$ becomes smaller than the natural diffraction divergence of the beam, the further approach of the incident wave to the total internal reflection angle becomes senseless. Instead of the energy concentrating in the reflected beam $r2$, the more and more fraction of the incident beam will be out of resonance. Thus, a small divergence of the both beams proves to be an important requirement which, in turn, limits a permissible sound frequency ν from below. Let us estimate these limitations.

In the case of total mode conversion, the condition for the balance of energy fluxes in the incident and reflected beams has the form $P_iD_i = P_{r2}d_{r2}$ (**Figure 1a**). This balance gives

$$K_2^{\rm con} \equiv P_{r2}/P_i = D_i/d_{r2}, \tag{26}$$

i.e., the reflected beam turns out to be narrower than the incident one by a factor of K_2. On the other hand, the related diffraction divergence angles, $\delta_i \sim c_s/\nu D_i$ and $\delta_{r2} \sim c_s/\nu d_{r2}$ (where $c_s \sim 10^5$ cm/s is the sound speed), are in similar proportion:

$$\delta_{r2} \approx \frac{D_i}{d_{r2}} \delta_i = K_2^{\rm con} \delta_i. \tag{27}$$

Thus, the possible increase in the coefficient K_2 is limited by the diffraction divergence of the $r2$ beam. To decrease this divergence, the frequency ν must be high. The simple estimation gives the following characteristic values: at $\nu \sim 100$ MHz and $D \sim 1$ cm, one can obtain a coefficient $K_2 \sim 5$–10 at $d_{r2} \approx 1$–2 mm, $\delta_i \sim 10^{-3}$, and $\delta_{r2} \sim 10^{-2}$ rad.

For a fixed direction of the normal $\mathbf{n}(\psi, \chi)$ to the crystal boundary, the surfaces $K_2(\varphi, \delta\alpha)$ and $\eta(\varphi, \delta\alpha)$ have specific "ridges" (**Figure 3**). They are determined by the special extremal relations between the angles φ and $\delta\alpha$. In framework of our approximate theory [8] (see Section 6), these trajectories coincide and are described by the relation of the type $\delta\alpha = C\varphi^4$. And along the ridges, one obtains

$$K_2^{\rm max} = {\rm const}/\sqrt{\delta\alpha}, \quad \eta^{\rm max} = {\rm const}. \tag{28}$$

Of course, with variations of $\mathbf{n}$, the constants in (28) also change. For some definite direction of $\mathbf{n}$, the mode conversion occurs when $\eta^{\rm max} = \eta^{\rm con} = 1$ and $K_2^{\rm max} = K_2^{\rm con}$.

Unfortunately, the obtained identity of the extremal trajectories is just the consequence of our approximations. As numerical analysis shows, usually they are close but not identical. And on one of them, we deal with the mode conversion, $\eta^{\rm con} = 1$ and $K_2^{\rm con}$, whereas on the other, an increased extremal gain $K_2^{\rm max} > K_2^{\rm con}$ and a decreased efficiency $\eta^{\rm max} < 1$ occur. It is even more important that on the second trajectory, the same value of coefficient K_2 is obtained at a larger angle $\delta\alpha$. And the numerical analysis shows that one can significantly increase $\delta\alpha$ in this case at the expense of a relatively small decrease in the efficiency η.

In **Figure 3** and in further considerations, we illustrate the performed analysis by numerical calculations for the series of crystals of various symmetry systems. The parameters of the resonance are calculated by exact formulas (5) and (8). We also vary the angle χ and do not limit ourselves to small angles φ and ψ.

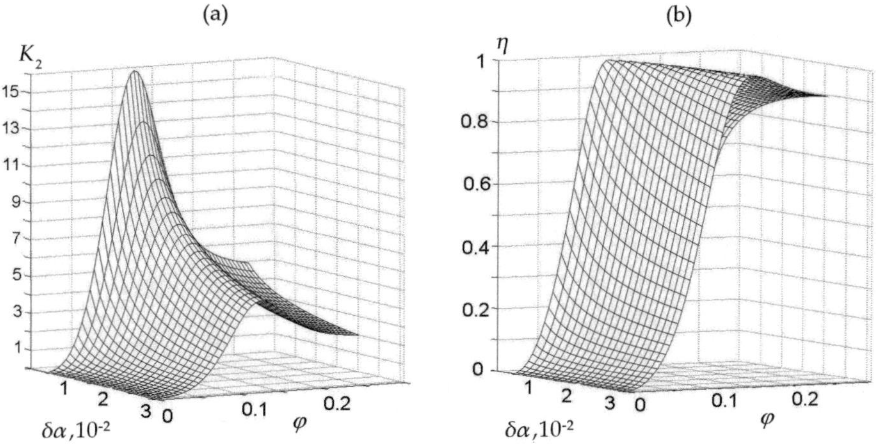

Figure 3.
Numerical plot of surfaces $K_2(\varphi, \delta\alpha)$ (a) and $\eta(\varphi, \delta\alpha)$ (b) for graphite crystal with the fixed boundary parallel to the hexad axis 6; the φ angle is counted from the vector $\mathbf{m}_0 \perp 6$.

Systems	Crystal	$\mathbf{m}_0$, $\mathbf{n}_0$	Normal n		φ_m, rad	$\delta\alpha_m$, rad	η, %
			χ, rad	ψ, rad			
Cubic	LiF	$\mathbf{m}_0 \| x', \mathbf{n}_0 \| y'$	0.21	0.29	0.13	0.008	88.7
	Silicon		0.29	0.29	0.13	0.007	74.0
Hexagonal	Graphite	$\mathbf{m}_0 \| x, \mathbf{n}_0 \| y$	0	0.02	0.14	0.034	94.9
	ZnS		0.03	0.48	0.32	0.008	93.4
	CdCe		0.03	0.69	0.14	0.008	95.2
Tetragonal	Paratellurite	$\mathbf{m}_0 \| x', \mathbf{n}_0 \| y'$	0.09	0.03	0.07	0.013	99.8
	$BaTiO_3$	$\mathbf{m}_0 \| x, \mathbf{n}_0 \| z$	0.05	0.28	0.37	0.031	80.0
Trigonal	Quartz	$\mathbf{m}_0 \| x_1, \mathbf{n}_0 \| y$	0.01	0.60	0.48	0.043	75.0
	$LiNbO_3$	$\mathbf{m}_0 \| x_2, \mathbf{n}_0 \| y$	0.30	0.78	0.09	0.007	79.0
Orthorhombic	Rochelle salt	$\mathbf{m}_0 \| z, \mathbf{n}_0 \| x$	0.12	0.44	0.55	0.009	94.6
		$\mathbf{m}_0 \| x, \mathbf{n}_0 \| z$	0.18	0.47	−0.08	0.010	98.4

The unperturbed orientations of vectors $\mathbf{m}_0$ and $\mathbf{n}_0$ are determined by the axes x, y, and z of crystallographic coordinates, the bisector axes x′ and y′ in the basal plane xy *and by the directions $x_\alpha \| (\cos\theta_\alpha, 0, \sin\theta_\alpha)$, where $\theta_1 = -0.76$ and $\theta_2 = 0.46$ rad.*

Table 1.
Characteristics of extremal resonances $K_2 = 5$ for acoustic crystals of various symmetry systems.

Table 1 demonstrates the results of such analysis for a number of acoustic crystals. We present the geometries related to the extremal gains. The angles $\varphi_m, \psi_m, \chi_m$ and $\delta\alpha_m$ are found for the gains $K_2^{\max} = 5$. In the presented examples, the compromise choice of the extremal geometry, instead of mode conversion one, indeed leads to a substantial increase in adjusting angles $\delta\alpha_m$ with efficiency η_m retained rather high level. The magnitudes of $\delta\alpha_m$ remain fairly small ($\sim$0.01 rad on average) even after the compromise but still look to be acceptable for an experiment. On the other hand, for quartz, graphite, and $BaTiO_3$ crystals, the angle $\delta\alpha_m$ is several times larger than the mentioned mean values. In the case of graphite, the mentioned compromise leads to the increase of the tuning angle by a factor of 1.5 via reducing the efficiency η by only 5%.

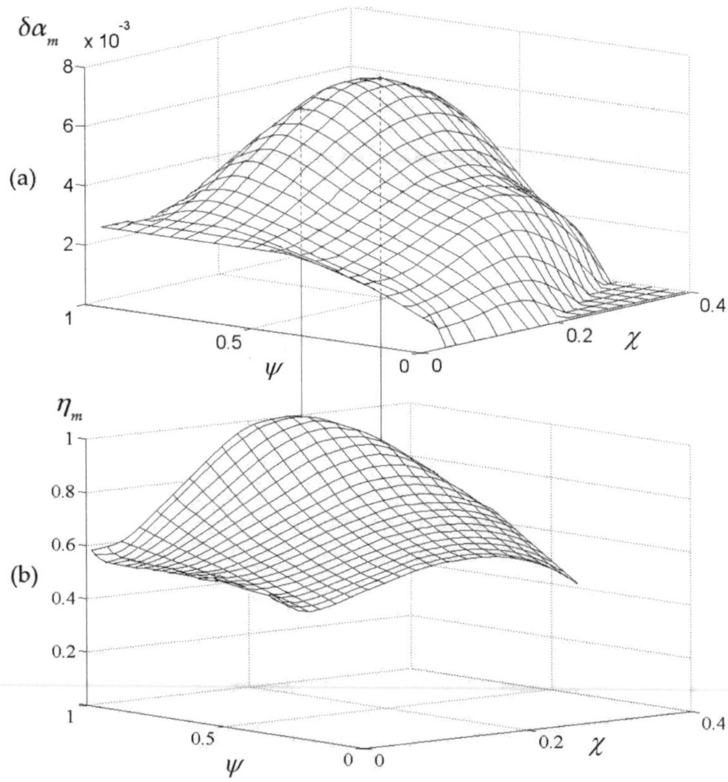

Figure 4.
Angle of incidence $\delta\alpha_m$ and efficiency η_m versus the angles ψ and χ at K_2 = 5 for lithium niobate ($LiNbO_3$) crystal at $\mathbf{m}_0 \| x_2$, $\mathbf{n}_0 \| y$, x_2 = ($\cos\theta_2$, 0, $\sin\theta_2$), and θ_2 = 0.46.

Figure 4 shows the optimization of the parameters $\delta\alpha_m$ and η_m for the lithium niobate crystal that corresponds to the value K_2 = 5 at the variation of the orientations of the normal **n** to the surface, i.e., the angles ψ and χ. For each **n** direction, the surface $K_2(\varphi, \delta\alpha)$ similar to that shown in **Figure 3** was plotted from which the φ_m and $\delta\alpha_m$ values corresponding to the extremal point at the "crest" with the amplitude K_2 = 5 were determined. Consequently, each point on the $\delta\alpha_m(\psi,\chi)$ and $\eta_m(\psi,\chi)$ surfaces in **Figure 4** corresponds to a certain angle φ_m. As is seen in the figure, the variations of the angles ψ and χ can significantly increase $\delta\alpha_m$ and η_m.

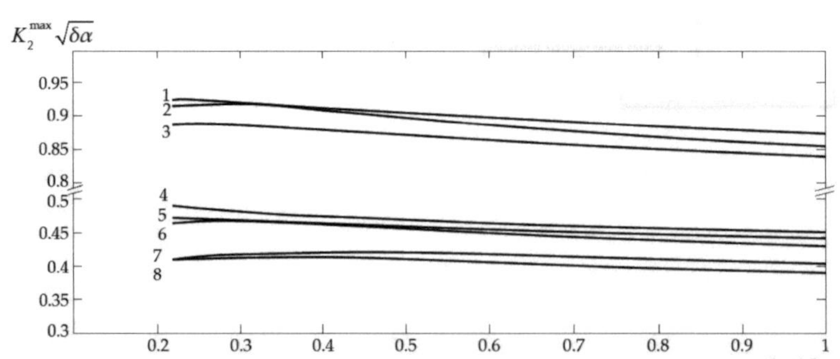

Figure 5.
Numerical plot of the product $K_2^{max}\sqrt{\delta\alpha}$ versus $\delta\alpha$ for the series of acoustic crystals (1—$BaTiO_3$, 2—paratellurite, 3—graphite, 4—CdCe, 5—ZnS, 6—LiF, 7—$LiNbO_3$, 8—Si).

The positions of maxima on the surfaces do not coincide but are quite close to each other, which allows a reasonable compromise at the choice of geometry.

Let us now check to what extent the found relations between parameters of our resonance reflection found in the first order of the perturbation theory retain their validity in a more precise numerical description. To be exact, we are checking the analytical dependence (28) of the extremal gain on the tuning angle $\delta\alpha$. And indeed, the numerical plot in **Figure 5** of the products $K_2^{\max}\sqrt{\delta\alpha}$ versus $\delta\alpha$ for eight analyzed acoustic crystals shows these dependences as very slowly changing functions close to constants.

6. Explicit theory for hexagonal crystals

In this section based on [11], we present the explicit analytical theory of the effect under consideration for a hexagonal medium with transversely isotropic elastic properties, which makes it possible to specify the above general relations and express the geometric conditions for mode conversion in terms of the moduli of elasticity of crystal. Analytical considerations will be supplemented with numerical calculations for some of hexagonal crystals.

To describe a hexagonal crystal, we use a standard crystallographic system of coordinates with the z axis oriented parallel to principal symmetry axis 6 and the x and y axes orthogonal to the z axis and lying in the basal plane of transverse isotropy [14, 15]. We choose the crystal boundary P_0 to be parallel to the axis 6 so that the normal $\mathbf{n}_0$ to this boundary is directed along the y axis. In this geometry, an EBW with the polarization $\mathbf{A}_{02}||z$ can propagate along the crystal surface in the direction $\mathbf{m}_0||x$ with a speed:

$$\hat{v}_0 = \sqrt{c_{44}/\rho}. \tag{29}$$

For transverse isotropy, we may change the initial crystal surface orientation $P_0 \to P$, rotating its normal vector $\mathbf{n}$ around $\mathbf{m}_0$ (i.e., choosing in **Figure 2a** the angle $\chi = \pi/2$) by a small angle ψ

$$\mathbf{n}_0 \to \mathbf{n} = (0,\ \cos\psi,\ \sin\psi). \tag{30}$$

In addition, as before, we introduce a perturbed propagation direction $\mathbf{m}$ rotated relative to the vector $\mathbf{m}_0$ by a small angle ϕ in the new surface plane P:

$$\begin{gathered}\mathbf{m}_0 \to \mathbf{m} = \mathbf{m}_0 \cos\varphi + [\mathbf{n}\times\mathbf{m}_0]\sin\varphi,\\ [\mathbf{n}\times\mathbf{m}_0] = (0,\ \sin\psi, -\cos\psi).\end{gathered} \tag{31}$$

Based on the standard equations of crystal acoustics [14, 15], one can determine the wave parameters entering superposition (1)

$$\begin{aligned}&\mathbf{A}_{1,4} = (\pm p, 1, 0)\sqrt{c_{66}/c_{44}},\\ &\mathbf{L}_{1,4} = \{2c_{66} - c_{44}, -2c_{66}p, (\pm p\psi - \varphi)c_{44}\}\sqrt{c_{66}/c_{44}},\\ &\mathbf{k}_{1,4} = k(1, \pm p, -\varphi);\end{aligned} \tag{32}$$

$$\begin{aligned}&\mathbf{A}_2 = (\varphi d/\Delta_{14}, 0, 1),\\ &\mathbf{L}_2 = \{\psi, \beta\varphi, \delta p + \tilde{\beta}\varphi\psi\}c_{44},\\ &\mathbf{k}_2 = k\left(1 - \varphi^2/2, \delta p + \varphi\psi, -\varphi\right);\end{aligned} \tag{33}$$

$$\mathbf{A}_3 = \{1, iq, (-\varphi + iq\psi)d/\Delta_{14}\}\sqrt{c_{11}/c_{44}},$$
$$\mathbf{L}_3 = \{2ic_{66}q, c_{44} - 2c_{66}, -c_{44}(\psi + i\kappa q\varphi)\}\sqrt{c_{11}/c_{44}}, \tag{34}$$
$$\mathbf{k}_3 = k(1, iq, -\varphi).$$

The following designations are introduced in formulas (32)–(34):

$$p = \sqrt{\Delta_{46}/c_{66}}, \quad q = \sqrt{\Delta_{14}/c_{11}}; \tag{35}$$

$$\beta = \frac{c_{12}d - \Delta_{13}\Delta_{14}}{c_{44}\Delta_{14}}, \quad \tilde{\beta} = \frac{c_{13}d - \Delta_{34}\Delta_{14}}{c_{44}\Delta_{14}}, \quad \kappa = 1 + \frac{d}{\Delta_{14}}; \tag{36}$$

$$\Delta_{ij} = c_{ii} - c_{jj}, \quad d = c_{44} + c_{13}. \tag{37}$$

We assume the parameters q and p to be real, which holds true at $c_{11} > c_{44} > c_{66}$. Note that the inequality $c_{11} > c_{66}$ is always satisfied (this is the crystal stability condition [14]) and the inequality $c_{11} > c_{44}$ is almost always satisfied (we do not know exclusions). Meanwhile, the condition $c_{44} > c_{66}$, which indicates that EBW (29) belongs to the middle sheet of the slowness surface (**Figure 2b**), is valid in far from all crystals (say, in a half of them).

As before, the angle of incidence is chosen near the angle $\hat{\alpha}$ of total internal reflection (**Figure 2b**). This angle corresponds to the limiting speed, which now can be directly related to the EBW speed $\hat{v}_0$ (29):

$$\hat{v}^2 = \hat{v}_0^2\left(1 - \widehat{\beta}\,\varphi^2\right), \qquad \widehat{\beta} = \frac{d^2 - \Delta_{14}\Delta_{34}}{c_{14}\Delta_{14}}. \tag{38}$$

In turn, the small tuning angle $\delta\alpha$ corresponds to the interval $\delta v = v - \hat{v}$ and, consequently, to the parameter δp:

$$\delta v = \hat{v}_0 p\delta\alpha = \frac{1}{2}\hat{v}_0\delta p^2, \qquad \delta\alpha = \delta p^2/2p. \tag{39}$$

Substituting expressions (32)–(34) for the vectors $\mathbf{L}_\alpha$ into Eq. (5), we obtain reflectances R_1 and R_2 as functions of the moduli of elasticity and perturbation parameters ϕ, ψ, and δp:

$$R_1 = \frac{\varphi^2 + ia\psi^2 - (\lambda' + i\lambda'')\left(\delta p + \widehat{\beta}\,\varphi\psi\right)}{-\varphi^2 + ia\psi^2 - (\lambda' - i\lambda'')\left(\delta p + \widehat{\beta}\,\varphi\psi\right)}, \tag{40}$$

$$R_2 = \frac{\mu\varphi + i\tilde{\mu}\psi}{\varphi^2 - ia\psi^2 + (\lambda' - i\lambda'')\left(\delta p + \widehat{\beta}\,\varphi\psi\right)}, \tag{41}$$

where

$$a = \frac{p}{q\beta^2}, \quad \lambda' = p\left(\frac{2c_{66}}{\beta c_{44}}\right)^2, \quad \lambda'' = \frac{1}{q}\left(\frac{2c_{66} - c_{44}}{\beta c_{44}}\right)^2,$$
$$\mu = \frac{4pc_{66}}{\beta c_{44}}\sqrt{\frac{c_{66}}{c_{44}}}, \quad \tilde{\mu} = \frac{2p(2c_{66} - c_{44})}{q\beta^2 c_{44}}\sqrt{\frac{c_{66}}{c_{44}}}. \tag{42}$$

Taking into account Eq. (40), we can reduce the conversion condition $R_1 = 0$ to the system of equations:

$$\begin{aligned} &\lambda''\varphi^2 - a\lambda'\psi^2 = 0, \\ &\varphi^2 - \lambda'\left(\delta p + \varphi\psi\,\widehat{\beta}\right) = 0. \end{aligned} \tag{43}$$

The first equation yields two versions of the mode conversion relationship between the rotation angles of the boundary (ψ) and sagittal plane (ϕ):

$$\psi_{\text{con}} = \pm\gamma\varphi_{\text{con}}, \qquad \gamma = \sqrt{\frac{\lambda''}{a\lambda'}}. \tag{44}$$

Substituting the parameter values from Eq. (42) into the radicand of Eq. (44), we can easily see that for $c_{44} > c_{66}$ in the case under consideration, the γ value is always real:

$$\gamma = \left|\frac{(c_{44} - 2c_{66})[c_{44}(c_{11} - 2c_{66}) + c_{13}(c_{44} - 2c_{66})]}{2c_{44}(c_{11} - c_{44})\sqrt{c_{66}(c_{44} - c_{66})}}\right|. \tag{45}$$

In other words, at $c_{44} > c_{66}$, there are always two crystal cuts, i.e., two versions of coupled orientations of the boundary and sagittal planes, which ensure mode conversion near angle $\hat{\alpha}$ of total internal reflection. Certainly, each set of chosen angles $\psi = \pm\gamma\varphi$ corresponds to its own definite angle of incidence: $\delta\alpha = \delta p^2/2p$ (39). The second equation in (43) yields:

$$\delta p_{\text{con}}^{\pm} = \varphi_{\text{con}}^2\left(\frac{1}{\lambda'} \mp \gamma\,\widehat{\beta}\right), \qquad \delta\alpha_{\text{con}}^{\pm} = \frac{\varphi_{\text{con}}^4}{2p}\left(\frac{1}{\lambda'} \mp \gamma\,\widehat{\beta}\right)^2. \tag{46}$$

Thus, the resonance width with respect to the incidence angle is indeed very small: $\delta\alpha \propto \varphi^4$. As we have seen, it is preferable to choose the sign corresponding to the maximum value $\delta\alpha_{\text{con}}^{\pm}$ (the upper sign at $\widehat{\beta} < 0$ and the lower sign at $\widehat{\beta} > 0$):

$$\psi_{\text{con}} = -\gamma\varphi_{\text{con}}\,\text{sgn}\,\widehat{\beta}, \qquad \delta\alpha_{\text{con}} = \frac{\varphi_{\text{con}}^4}{2p}\left(\frac{1}{\lambda'} + \gamma\left|\widehat{\beta}\right|\right)^2. \tag{47}$$

Given the found relations for reflectances (40) and (41), we can obtain the gain (K_2) and loss (K_1) coefficients (8) where one can put

$$s_1 \approx s_4, \qquad s_2/s_4 \approx \sqrt{c_{44}/c_{66}}. \tag{48}$$

Here we took into account that for the studied reflection geometry with the sagittal plane close to transverse isotropy, the ray speeds s_α are approximately equal to the phase speeds of the bulk waves involved in the reflection. Under the total conversion conditions (44)–(47), the coefficient K_1 is 0; therefore, the excitation efficiency of beam $r2$ is maximum: $\eta = 1 = 100\%$. In this case, the gain K_2 can be written as:

$$K_2^{\text{con}} = \frac{G_{\text{con}}}{\varphi_{\text{con}}^2}, \qquad G_{\text{con}} = \frac{(\mu^2 + \tilde{\mu}^2\gamma^2)s_2/s_4}{4(1 + a^2\gamma^4)}. \tag{49}$$

The above analysis based on expansion of the equations in small parameters φ^2, ψ^2, δp, and $\delta\alpha$ is approximate. Therefore, the range of applicability of the results obtained may differ, depending on the degree of crystal anisotropy and other factors. In **Figure 6**, the analytical linear relation between the conversion angles φ

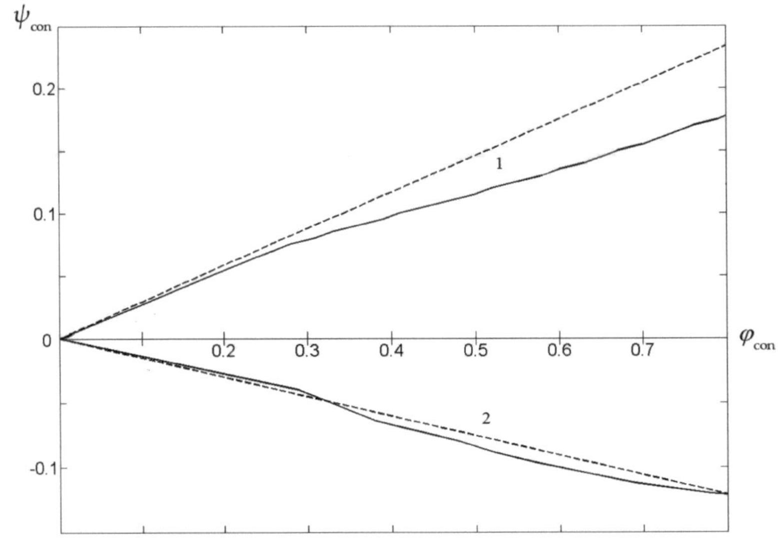

Figure 6.
Correlations between the conversion angles ψ_{con} and φ_{con} (in radians) obtained by numerical calculation (solid lines) and approximate theoretical analysis (dashed lines) for the (1) Ti and (2) BeCu crystals.

and ψ (47) is compared with the results of numerical calculations based on formulas (5), without their expansion in small parameters, on the example of two (Ti, and BeCu) hexagonal crystals [16, 17].

In both cases, our computations not only practically confirm linearity of the relation between ϕ and ψ but also yield slopes of these dependences close to theoretical ones. In case of Ti, we have $\overset{\frown}{\beta} < 0$ and coincident signs of angles ϕ and ψ, while for BeCu, $\overset{\frown}{\beta} > 0$ and the angle signs are different.

We can only wonder why the predictions in the first order of the perturbation theory are confirmed so well by the exact numerical calculation in a wide range of angles ϕ, which are far from small. Anyhow, but the Ti crystal reveals one more "loyalty" with respect to our approximate theory: in its case, the geometries of the mode conversion and the extremal gain are almost identical.

7. The particular case of 1D perturbation (ψ = 0)

The presented theory of mode conversion in the vicinity of the angle of total internal reflection was based on coupled perturbations of both the sagittal plane and the surface orientations. At the rotations of only one of these planes, the mode conversion can be achieved only for crystals with certain relations between the moduli of elasticity. However, sometimes, these requirements might appear to be not very limiting. This can be explicitly shown on the example of the considered above case of hexagonal crystals. Indeed, for the unperturbed surface at $\psi = 0$, the loss coefficient K_1 (40) acquires a minimal value K_1^{min} at $\varphi^2 = \lambda' \delta p$ which may be estimated as:

$$K_1^{\min} = |R_1|^2 = \left|\frac{\lambda''}{2\lambda'}\right|^2 \sim \left(\frac{c_{44} - 2c_{66}}{c_{44}}\right)^4. \tag{50}$$

Thus, any hexagonal crystal with the modulus c_{44} close to $2c_{66}$ must be very efficient for our effect ($K_1 \ll 1$). Due to the "4" power in the latter estimate, this remains true even when the moduli are not so close to each other. For instance, at $|c_{44}-2c_{66}|/c_{44} \sim 1/3$, Eq. (50) gives the estimate $K_1 \sim 10^{-2}$ and efficiency $\eta = 1-K_1$ becomes $\sim 99\%$. In accordance with reference book [16], there are quite a number of hexagonal crystals where $K_1 \ll 1$. Below, we shall also give examples of crystals of monoclinic symmetry systems having got the same property.

This motivates us to a short consideration of the simplified approach to the reflection resonance with unchanged crystal surface orientation. As was shown in [8], the description becomes especially compact if to choose the crystal boundary parallel to the plane of crystal symmetry. By the way, such planes exist in all crystals, except triclinic [14, 15]. In this case, expressions (21) and (22) acquire the structure:

$$K_1 = \frac{(\varphi^2 - \lambda'\delta p)^2 + (\lambda''\delta p)^2}{(\varphi^2 + \lambda'\delta p)^2 + (\lambda''\delta p)^2}, \tag{51}$$

$$K_2 = \frac{(\mu\varphi)^2 s_2/s_4}{(\varphi^2 + \lambda'\delta p)^2 + (\lambda''\delta p)^2}. \tag{52}$$

And the efficiency of the resonance $\eta = 1-K_1$ is given by:

$$\eta = \frac{4\lambda'\delta p\varphi^2}{(\varphi^2 + \lambda'\delta p)^2 + (\lambda''\delta p)^2}. \tag{53}$$

Thus, for a fixed δp, the coefficients K_2 and η are determined by the same function $F(\varphi)$:

$$K_2(\varphi) = \frac{\mu s_2}{s_4}F(\varphi), \quad \eta(\varphi) = 4\lambda'\delta p F(\varphi), \quad F(\varphi) = \frac{\varphi^2}{(\varphi^2 + \lambda'\delta p)^2 + (\lambda''\delta p)^2}. \tag{54}$$

Accordingly, their maximum magnitudes are determined by the same extremum condition:

$$\varphi^2 = |\lambda|\delta p = \delta p\sqrt{\lambda'^2 + \lambda''^2} \approx \lambda'\delta p\left[1 + \frac{1}{2}\left(\frac{\lambda''}{\lambda'}\right)^2\right]. \tag{55}$$

The last estimate in (55) is valid when $(\lambda''/\lambda')^2 \ll 1$. In this approximation, one obtains

$$K_1^{\min}(\varphi, \delta\alpha) \approx \left(\frac{\lambda''}{2\lambda'}\right)^2, \quad K_2^{\max}(\varphi, \delta\alpha) \approx \frac{G}{\varphi^2}, \quad G \approx \frac{\mu s_2}{s_4}. \tag{56}$$

Thus, one can conclude that the consistent variation of the sagittal plane orientation φ and the tuning incidence angle $\delta\alpha \propto \delta p^2$ (10) along one extremal trajectory (55) provides simultaneous optimization of both the gain and the efficiency of the resonance reflection. However, as was shown above, this coincidence is not an exact result but a consequence of our approximate calculations. The computer analysis based on exact formulas (5) without their expansion in small parameters leads to distinct extremal trajectories for the functions $K_2^{\max}(\varphi, \delta\alpha)$ and $\eta^{\max}(\varphi, \delta\alpha)$. A difference between them depends on angle φ and crystal anisotropy. On the other hand, as was discussed above, the occurrence of those trajectories might be used for substantial increasing of the width of the resonance in the range of tuning incidence angle $\delta\alpha$ at the expense of a slight decrease in the efficiency η.

Figure 7 shows dependences $K_2(\delta\alpha)$ and $\eta(\delta\alpha)$ in such extremal trajectories related to ridges on the surfaces $K_2(\varphi, \delta\alpha)$ and $\eta(\varphi, \delta\alpha)$ (of the type shown in **Figure 3**) conformably to a monoclinic stilbene crystal. The upper curves correspond to the choice of $\varphi = f_{K,\eta}(\delta\alpha)$ in the "proper" trajectory corresponding to the maximum of the shown characteristic, while the lower curves are plotted for $\varphi = f_{\eta,K}(\delta\alpha)$ from the "foreign" trajectory. In this case, of the two possible optimization variants, the trajectory in which $K_2^{\max}$ is realized is more advantageous. This corresponds to curves *1* in **Figure** 7. Obviously, for such a choice in the stilbene case, there is approximately only a 2.5%, loss in efficiency, but in the return gain, twice in the resonance width: the value of $\delta\alpha_{\mathrm{m}}$ for the optimal angle of incidence corresponding to an amplitude of $K_2^{\max} = 5$ increases from 0.08 to 0.155 rad.

In [8], we accomplished the numerical search of crystal candidates for possible future observations of the discussed effect. In all cases, the surface was supposed to be parallel to the symmetry plane while the sagittal plane orientation varied. The "casting" involved about 350 crystals. The basic criteria for the crystal selection were the closeness of the resonance to mode conversion and not too small resonance widths over the angles of incidence ($\delta\alpha_m \geq 0.01$ rad for $K_2 = 5$). According to such criteria, we found 14 crystals of monoclinic, trigonal, orthorhombic, and hexagonal systems. They are characterized by a rather high efficiency η_m, while the width $\delta\alpha$ of resonances over the angles of incidence satisfies the formulated selection rule. The resulting parameters of the resonance in these crystals are described in [8].

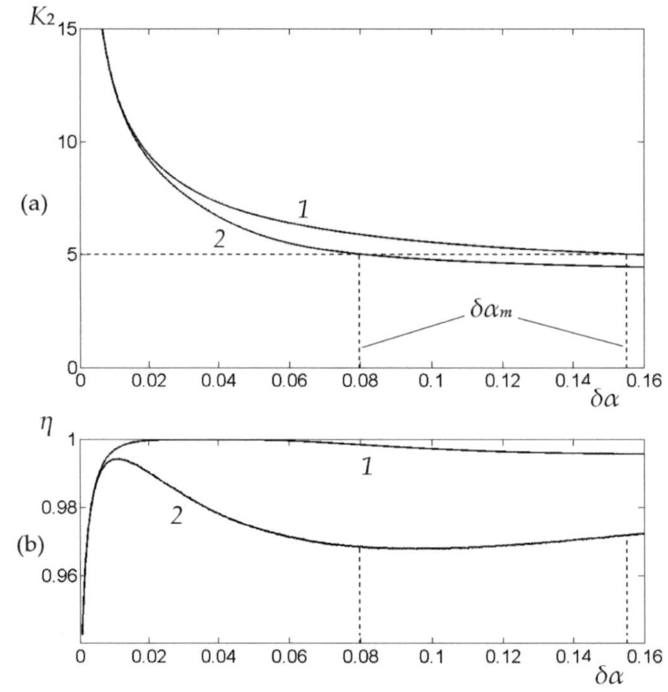

Figure 7.
Dependences of the gain coefficient K_2 (a) and the efficiency η (b) of the resonance on the incidence angle $\delta\alpha$ in a stilbene crystal with the surface parallel to the symmetry plane for optimal trajectories corresponding to the "ridges" of surfaces $K_2(\varphi, \delta\alpha)$ (curves 1) and $\eta(\varphi, \delta\alpha)$ (curves 2).

Crystals	φ_0, rad	φ_m, rad	$\delta\alpha_m$, rad	η_m, %
Stilbene	1.54	−0.16	0.155	97.2
Triglycine sulfate	1.35	0.12	0.089	91.0
Benzyl	1.03	0.73	0.069	97.6
Tartaric acid	−1.13	−0.30	0.021	97.4

The directions of $\mathbf{m}_0$ are specified by the azimuth angles φ_0 counted from the x axis of the crystallographic coordinates and the angles φ_m—From $\mathbf{m}_0$.

Table 2.
Key parameters of 1D resonance for a series of monoclinic crystals.

Here, we limit ourselves to presenting only data for several monoclinic crystals with the best parameters (see **Table 2**). The values of $\delta\alpha_m$, φ_m, and η_m presented in the table for all crystals correspond to the same gain $K_2 = 5$ and relate to the choice of optimal trajectory φ_K as for stilbene (**Figure 7**). This results in the maximum width of resonance for a small decrease in efficiency η_m. By the way, stilbene is the absolute leader in the table in the value of $\delta\alpha_m$ and one of the leaders in the efficiency η_m.

8. Conclusions

The analytical theory developed above is constructed within the theory of elasticity and, within the range of its applicability, it is exact to the extent of Eqs. (5) and (8), which express the resonant reflection coefficients for an arbitrary anisotropic medium in terms of the eigenvectors of Stroh's matrix (3). Generally, the dependence of these eigenvectors on the geometrical parameters of reflection can easily be found by numerical methods. An analytical alternative is to expand the exact formulas (5) into a series in small angular parameters. Finally, an explicit analytical calculation based on these formulas for a number of geometries in high-symmetry crystals (for example, hexagonal ones) is also possible. Here, we used all three approaches: the numerical calculations based on Eqs. (5) and (8), their expansion into a series, and even an explicit representation of the results via the elastic moduli. In this case, avoiding the cumbersomeness of our calculations and the unmanageability of the analytical formulas, we retained only the first non-vanishing terms in all expansions that conveyed the key dependences and the effect being investigated on physical parameters. On the other hand, all graphical results of our analysis were obtained through computations based on the exact formulas (5) and (8).

Based on the same principles, we used the image of acoustic beams in our reasoning only for clarity. Actually, we did not go outside the plane wave approximation in our calculations by assuming it to be sufficient in the short wavelength limit of interest, $\lambda/D \sim 10^{-3}$ rad. Here, we also had in mind the possible manifestations of the effect in phonon physics, where the language of plane waves is more relevant.

Based on our analysis, we can probably count on the realization of resonant reflection in crystals, whereby a wide incident acoustic beam converts almost all of its energy into a narrow high-intensity reflected beam. A special choice of crystals with a definite relation between the elastic moduli is required to optimize the resonance. In addition, since the resonance region is narrow in angles of incidence, stringent requirements for a weak divergence of the incident beam, $\sim 10^{-3}$ rad, which can be realized only at high ultrasonic frequencies ~100 MHz, arise. For the same reason, the amplitude of the excitation coefficient is also limited to $K_2 \approx 5$–10. However, in the case of retransformation of the emergent beam through its narrowing in the perpendicular dimension as well, the intensification efficiency increases many fold, to $\sim 10^2$. In the hypersonic frequency range, the amplification amplitudes can be increased significantly. In this case, however, one might expect additional restrictions due to an increase in the absorption of acoustic waves.

The acoustic resonance considered here can also manifest itself in phonon physics as the channeling of high-density energy near the surfaces of crystals with specially chosen orientations and in accompanying nonlinear phenomena. The strategy for an experimental search of such effects can be based on existing techniques for studying the manifestations of phonon focusing [3] and the propagation of ballistic phonons [18].

Acknowledgements

This work was supported in part by the Ministry of Education and Science of the Russian Federation (state assignment for Federal Research Center "Crystallography and Photonics").

Author details

Vladimir I. Alshits*, Dmitrii A. Bessonov and Vasilii N. Lyubimov
A.V. Shubnikov Institute of Crystallography, Federal Research Center "Crystallography and Photonics," Russian Academy of Sciences, Moscow, Russia

*Address all correspondence to: valshits@mail.ru

References

[1] Royer D, Dieulesaint E. Elastic Waves in Solids. Vol. 1, 2. Berlin: Springer; 2000. DOI: 10.1007/978-3-642-02589-1

[2] Aleksandrov KS, Sorokin BP, Burkov SI. Effective Piezoelectric Crystals for Acoustoelectronics, Piezotechnics and Sensors. Vol. 1, 2. Novosibirsk: Sib. Otd. RAN; 2007-2008

[3] Wolfe JP. Imaging Phonons. Cambridge, UK: Cambridge University Press; 2005

[4] Ponomarev AE, Bulatitskii SI, Sapozhnikov OA. Compression and amplification of an ultrasonic pulse reflected from a one-dimensional layered structure. Acoustical Physics. 2007;**53**:127-135. DOI: 10.1134/S1063771007020030

[5] Alshits VI, Bessonov DA, Lyubimov VN. Resonant excitation of intense acoustic waves in crystals. Journal of Experimental and Theoretical Physics. 2013;**116**:928-944. DOI: 10.1134/S1063776113060010

[6] Chadwick P, Smith GP. Foundations of the theory of surface waves in anisotropic elastic materials. Advances in Applied Mechanics. 1977;**17**:303-376. DOI: 10.1016/S0065-21560870223-0

[7] Alshits VI, Lothe J. Elastic waves in triclinic crystals. I. General theory and the degeneracy problem. Soviet Physics Crystallography. 1979;**24**:387-392

[8] Alshits VI, Bessonov DA, Lyubimov VN. Resonance compression of an acoustic beam in a crystal. Journal of Experimental and Theoretical Physics. 2016;**122**:689-704. DOI: 10.1134/S1063776116020151

[9] Alshits VI, Bessonov DA, Lyubimov VN. Excitation of intense acoustic waves in hexagonal crystals. Crystallography Reports. 2013;**58**:867-876. DOI: 10.1134/S1063774513060047

[10] Alshits VI, Bessonov DA, Lyubimov VN. Intensification of acoustic beams in crystals at mode conversion near total internal reflection. JETP Letters. 2017; **106**:40-45. DOI: 10.1134/S0021364017130069

[11] Lyubimov VN, Bessonov DA, Alshits VI. Compression of acoustic beams at conversion reflection in hexagonal crystals. Crystallography Reports. 2018;**63**:621-626. DOI: 10.1134/S1063774518040168

[12] Stroh AN. Steady state problems in anisotropic elasticity. Journal of Mathematical Physics. 1962;**41**:77-103

[13] Barnett DM, Lothe J. Free surface (Rayleigh) waves in anisotropic elastic half-spaces: The surface impedance method. Proceedings of the Royal Society of London. 1985;**A402**:135-152

[14] Fedorov FI. Theory of Elastic Waves in Crystals. New York: Plenum Press; 1968 (Moscow: Nauka; 1965)

[15] Sirotin Yu I, Shaskolskaya MP. Fundamentals of Crystal Physics. Moscow: Mir; 1982

[16] Landolt HH, Börnstein R. In: Hellwege K-H, editor. Zahlenwerte und Funktionenaus Naturwissenschaften und Technik. 11th ed. Vol. Neue Serie, III. Berlin: Springer; 1979

[17] Blistanov AA, Bondarenko VS, Perelomova NV, et al. Acoustic Crystals: A Handbook. Moscow: Nauka; 1982

[18] Alshits VI, Ivanov SN, Soifer Ya M, Taranov AV, Khazanov EN. The observation of dislocation flutter-resonance in the temperature dependence of scattering of nonequilibrium phonons in LiF crystals. Soviet Physics—Solid State. 1989;**31**:1873-1879

Chapter 9

In Situ Detection of Leakages in Partition Elements through SONAH and Beamforming Techniques

José A. Ballesteros, Samuel Quintana and Marcos D. Fernandez

Abstract

Airborne sound insulation in buildings, whether in fixed partition elements, like partitions or party walls, or mobile elements, like doors or screens, is always related to the performance of the weakest element involved. In situ assessment of airborne sound insulation in building elements can be carried out by pressure techniques or sound intensity techniques. Sound pressure techniques are very quick but fail to discriminate the sound insulation contribution of each building element involved. Sound intensity techniques, on the other hand, allow to determine the sound transmission of each element and also to discriminate indirect transmissions up to a certain degree. In order to find areas with high sound transmission, such as leakages or weakened regions, a large number of measurements on the building element surface have to be performed. Moreover, the sound intensity technique is very time-consuming, because it is necessary to carry out the measurement in each grid point defined. This chapter describes the use of beamforming and SONAH techniques to detect areas with lower airborne sound insulation in a building element. These techniques unify the advantages of both, pressure and sound intensity techniques, allowing the quick visualization of leakages or weakened areas of different building elements.

Keywords: building, insulation, leakage, SONAH, beamforming

1. Introduction

Standard in situ methods to obtain airborne sound insulation in building elements are based on sound pressure measurements (ISO 16283-1) [1] and on sound intensity measurements (ISO 15186-2) [2].

The method described in ISO 16283-1 [1] states that the acoustic field generated at the emitting room must be diffuse, stationary, and spectrally flat, at least for the frequency range under consideration (100–3150 Hz). Acoustic field and reverberation time measurements are averaged in space and time to ensure low statistical spread.

The apparent sound reduction index (R') is one of the parameters used to express the acoustic behavior, and it stems from Eq. (1):

$$R' [dB] = \overline{L_1} - \overline{L_2} + 10 \log\frac{S}{A} \tag{1}$$

$$A[m^2] = \frac{0.16V}{T} \tag{2}$$

where

- $\overline{L_1}$ [dB] is the average sound pressure level at the emitting room.
- $\overline{L_2}$ [dB] is the average sound pressure level at the receiving room.
- S [m^2] is the total surface of the common partition element between both rooms.
- A [m^2] is the equivalent acoustic absorption area at the receiving room. It is obtained using the Sabine Eq. (2).
- V [m^3] is the volume of the receiving room.
- T [s] is the reverberation time in the receiving room.

The main difference in the measurement procedure ISO-15186-2 [2] regarding to ISO 16283-1 [1] is that reverberation time measurement is not required, and in the receiving room, the measured parameter is the acoustic intensity normal to the surface of the partition element being assessed, whether by scanning or by grid techniques, depending on the resolution required. In this case, the acoustic behavior is defined by the apparent sound reduction index by intensimetry R'_I, which is different from the index obtained by acoustic pressure and is calculated according to Eq. (3):

$$R'_I [dB] = \left[\overline{L_{p1}} - 6 + 10 \log\frac{S}{S_0}\right] - \left[\overline{L_{In}} + 10 \log\frac{S_M}{S_0}\right] \tag{3}$$

where

- $\overline{L_{p1}}$ [dB] is the average noise pressure level at the emitting room.
- S [m^2] is the surface of the partition element under study.
- $\overline{L_{In}}$ [dB] is the average sound intensity level, normal to the measuring surface/s in the receiving room.
- SM [m^2] is the total surface of the measuring surface/s.
- S0 (1 m^2) is the reference surface.

The procedure described in ISO-15186-2 [2] provides a partial assessment of the sound insulation of each surface present in the partition element. Nevertheless, the final evaluation of the acoustic behavior depends strongly on the resolution defined throughout the measurement procedure and is closely related to a high density of measuring points. The validity range of this technique is determined through the calculation and the monitoring of the so-called field indicators (F_2, F_3, F_4) according to ISO 9614-1 [3], which allow to ascertain whether the acoustic intensity measurement conditions fulfill or not the minimal requirements.

After the examination of the measurement procedures, it is noticed that the main advantages of the pressure method [1] are its standardization and the reduced time required to take the measurement; its main disadvantages are the incapability of detecting leaks and areas with a poor insulation level, and, additionally, it does not reject the possible indirect transmissions. On the contrary, the technique based on intensity measurements [2], identifies leakages or weakened areas; its resolution, however, depends on the density of points of the measuring grid and hence, on the measuring time for each test. In order to achieve intensity data with high resolution in terms of space and frequency, a highly dense (every 5–10 cm) point grid must be designed, which leads to long measuring times.

A new measurement procedure is proposed to unify the advantages of both methods mentioned above: quickness and detection of leaks. This new procedure is based on beamforming and accompanied by SONAH; its main aim is the identification of the areas with weak insulation in one shot measurement [4].

Beamforming is performed, in its simplest approach, through the delay-and-sum algorithm, which consist of the sum of the delayed signals from the array microphones with different delays in order to put all the signals in phase. With this, the signal of interest is reinforced against the noise and other signals propagating in other directions [5].

In the statistically optimal near-field acoustic holography (SONAH) method [6], the acoustic quantities on a mapping surface near the measurement surface are calculated by using a transfer matrix defined in such a way that all the propagating waves and a weighted set of evanescent waves are projected with optimal average accuracy. The main advantage of SONAH is the fact that it does not use the discrete spatial Fourier transform used in the classical NAH procedure. Therefore, undesirable spatial leakage effects are avoided [7].

SONAH [6] and beamforming [8] techniques use a measurement system based on specially designed geometrical configurations of microphone arrays [9], which help to understand the acoustic field at a given distance from the acoustic source, using various signal processing algorithms. These techniques are mainly used to locate acoustic sources [10].

Besides the different array geometries [11] and algorithms [5] used, the main differences in practical applications between SONAH and beamforming techniques reside in:

- The distance between the array and the acoustic field—SONAH technique requires near-field measurements close to the surface [6], whereas beamforming technique does the measurements in the far field [12].
- In the covered surface of study.
- In the use of references, which are mandatory in SONAH.

Using only one of these two methods—beamforming or SONAH—would not suffice to cover the whole frequency range of interest (100–3150 Hz), since beamforming shows poor resolution in low-frequency range and SONAH requires a large number of microphones to attain good resolutions at high frequency [13]. Therefore, a combination of both methods can be applied throughout the desired range.

2. Methodology

All the recommendations laid down in measurements based on pressure [1] and intensity [2] standards have been interrelated regarding the execution of the beamforming and SONAH measurements.

In the emitting room, two omnidirectional sound sources, one placed close to a corner generating white noise and the other one in the center of the room generating pink noise, are independently triggered and equalized to generate an acoustic field diffuse, stationary, and spectrally flat (90–95 dB approx.) with no tonalities over the desired frequency range.

At the receiving room measurement positions are defined based on the features of the surface under study, the instrumentation, and the technique used.

Beamforming technique states that the measuring distance between the array and the source, to cover the whole area of interest at the partition element, must be the minimum value required according to Eq. (4) [14]:

$$L = 1.15z \tag{4}$$

where

- z [m] is the measuring distance.
- L [m] is the length of the sound source.

Nevertheless, the bigger the distance of Beamforming measurement is, the lower the resolution will be, and therefore, the more difficult will be to locate the maximum sound radiation areas. The resolution can be calculated according to Eq. (5):

$$R = 1.22\left(\frac{z}{D}\right)\lambda \tag{5}$$

where

- R [m] is the resolution.
- $\lambda\ [m] = \frac{c\ \left[\frac{m}{s}\right]}{f\ [Hz]}$ is the wavelength.
- D [m] is the array diameter.

Taking this into account, in the measurements, distances from the partition element should be chosen in such a way that the array can be placed at a distance that allows to fully cover the surface under study, carefully combining the values for calculation distance, covered surface, and expected resolution.

When the measurements have been taken, to process the data, the calculation points are defined over a flat grid. This grid should have an x/y axis spacing as dense as possible in order to allow an accurate identification of leaks, fissures, or sealing defects.

Once defined the calculation grid, the simplest algorithm that can be used is the delay-and-sum one. With this algorithm, it is possible to define the minimum working frequency for a 30° opening angle as [12].

$$f_{min}\left(30^{\circ}\right) = \frac{c}{D} \tag{6}$$

with

- c the speed of sound (344 m/s)
- D [m] the array diameter

The highest frequency is defined according to the sidelobe level as

$$f_{max}\left(30^{\circ}\right) = \frac{4}{3} f_T, \tag{7}$$

being f_T [Hz] the sidelobe threshold frequency [12].

The frequency range obtained can be used for the case of the beamforming delay-and-sum algorithm and could be increased if other algorithms (such as minimum variance or Capon [15], clean-SC [16], DAMAS [17], beamforming through using eigenvalues and eigenvectors [18], MUSIC [19], or orthogonal beamforming [20]) are used.

In the case of SONAH measurements, the instrumentation setup is configured to carry out the measurements in the near field. The calculation grid is limited by the array dimensions (although we can take more measurements to synthesize a larger grid [21]). Depending on the array dimensions and the possibility of synthesizing a large grid, the measurements can be performed in two manners:

- Over the whole partition element
- Over smaller characteristic areas, made with different materials, where there might be suspicions fissures or leakages

The use of references is compulsory in SONAH. They are selected according to the features of the sound field at the receiver (microphone references) [22] or to the vibration level at the partition elements (accelerometer references) [14], which could be related to sound insulation by the pressure and intensity calculations obtained by these techniques.

For SONAH measurements, the working frequency range is determined by

$$f_{min}\ [Hz] = \frac{d}{2D} \tag{8}$$

and

$$f_{max}\ [Hz] = \frac{c}{2dx} \tag{9}$$

where

- c is the speed of sound (344 m/s).
- dx [m] is the average spacing between array microphone positions.
- D [m] is the diameter of the array.

Once the array-based measurements are performed and processed, results are displayed in color maps where the areas of the studied surface with higher pressure or intensity levels can be identified. These areas, therefore, corresponding to leaks, fissures, or sealing defects, are those with higher sound transmission.

2.1 Practical application of the method

An applied experiment has been carried out to check the correctness, validity, and practicability of this new methodology to quickly detect in situ acoustic leakages.

Specifically, in situ measurements were performed through multiple testing of various partition elements present in different buildings. Partition elements were chosen as a broad sample of the most common elements in use, both by their typology and materials [23]. Eight different partition elements were selected to validate the method, such as laminated plasterboard wall and different kinds of homogeneous or heterogeneous doors made of metal, wood, or glass. The assessment of those doors considered their placement onto either a plasterboard system partition element or a normal brick wall, with better sound insulation properties than the door itself; in all cases, walls with minimum indirect transmission conditions were selected (i.e., surrounding walls with heavy elements and T junctions, without ventilation grids nor cabling boxes) [24]. **Figure 1** displays all the partition elements under study. **Table 1** summarizes the main features of the performed tests, where

- *Ve* [m^3] and *Vr* [m^3] are the volume of emitting and receiving rooms, respectively.

- *Sse* [m^2] and *Ssr* [m^2] are the floor surface of emitting and receiving rooms, respectively.

- *ST* [m^2] is the total surface of partition element or common compartmentalization element.

To perform beamforming and SONAH measurements, an 18-microphone slice wheel array (**Figure 2**) was used. Ten-second time signals were recorded with the array linked to a data acquisition system with a sample frequency of 65,536 Hz [25] governed by a computer.

In the beamforming measurements performed, the distances varied between 2 and 5 m from the partition element. **Figure 3** shows the beamforming layout used to carry out the measurements.

To process the data, the delay-and-sum algorithm was used over a flat grid with a maximum x/y axis spacing of 0.05 m. With this algorithm and an array diameter of 35 cm, the minimum working frequency would be about 1 kHz, according to Eq. (6). Moreover, defining a threshold $T = -8\,dB$, as it is established by the manufacturer of the array for the delay-and-sum algorithm, and taking into account the array characteristics, applying Eq. (7), a $f_{max}(30°)$ about 6143 Hz is obtained [26]. Taking this into account, the frequency range was defined from the third octave band centered on 1 kHz to the one centered on 5 kHz.

In SONAH measurements, since the array is 35 cm in diameter, they were performed at a distance on 10 cm on areas where fissures could be observed, which can be those characteristic areas of the door made with different materials or because they had some particularities. **Figure 4** shows the SONAH layout used to carry out the measurements.

To process the data, a flat grid with a maximum x/y axis spacing of 0.01 m was used. As the average spacing between measurement points is 0.073 m and D is 0.35 m, and applying Eqs. (8) and (9), the minimum working frequency would

Figure 1.
Photographs of partition elements under study (descriptions of each Id. included in ***Table 1****).*

be 123 Hz and the maximum working frequency 2353 Hz [26]. Taking this into account, the measurements were taken from the third octave band centered on 160 Hz to the one centered on 1600 Hz.

Both beamforming and SONAH tests were post-processed considering also the option of including reference signals. In this way, five reference signals were used: two vibrating signals and three sound pressure signals. Since some partition elements are made of two different materials, it was decided to place an accelerometer centered on the surface of each material and three microphones in the array plane along a diagonal line across the assessed surface [27]: top left, center, and bottom right. **Figures 3** and **4** show the reference configuration in each case.

		Emitter		Receiver		Surface	R'$_w$ [dB] ISO 16283-1
Id.	Description	Ve [m^3]	Sse [m^2]	Vr [m^3]	Ssr [m^2]	ST [m^2]	
1	Plasterboard wall	519	181	219	75	16.7	35
2	*Glass door*	98	33	244	85	5.3	18
3	Anti-panic door	246	91	124	39	11.9	24
4	*Wood door-2*	67	25	144	54	12.7	28
5	Metal door	114	48	200	213	11.9	23
6	*Wood door*	200	80	350	546	15.0	26
7	Metal door with glass spyhole	204	82	280	62	8.0	21
8	*Wood door with glass window*	421	145	170	60	8.5	19

Table 1.
Features of testing environment.

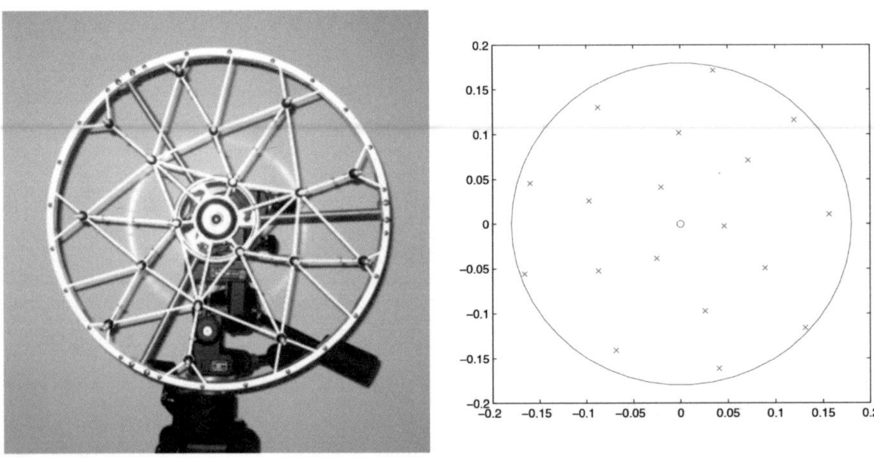

Figure 2.
Left, 18-microphone slice wheel array. Right, microphone positions (dimensions are given in meters).

The post-processed calculations with the reference signals were as follow:

- Without reference signals in the calculations (only in the beamforming case)
- Reference with the central microphone of the array
- Reference with three ½" microphones in the array plane along a diagonal line across the assessed surface: top left, center, and bottom right
- One accelerometer at the surface center in homogeneous elements or two accelerometers centered on the center of each material/surface for elements made of two different materials (e.g., wooden door with glass window—for instance **Figure 1**, Id. 8).

The best results were obtained in beamforming measurements without using references at all and using the central microphone of the array or accelerometers located in areas with different materials in SONAH measurements.

Figure 3.
Example of the beamforming measurement layout.

As the amount of data collected is very large, only the results for the base partition element and a typical and representative example for one of the doors (anti-panic door, **Figure 1**; Id. 3) are exposed. The remaining elements under study were treated in a similar way.

In order to compare the results obtained with those from a standardized technique, the partition elements under test were also measured using the intensity standard described in ISO 15186-2 [2]. The configuration of the emitting room was the same than in the array-based measurements. In the receiving room, measurement grids with at least one point every 30 cm in horizontal and 40 cm in vertical were defined, carrying out two measurement sequences and averaging in each measurement point the intensity obtained for both measurements. Twenty-second measurements were taken in each measurement point using the 12-mm separator, which make possible the measure in a frequency range from the third octave band centered on 200 Hz to the one centered on 5 kHz. Moreover, the calculation of field indicators (F_2, F_3, F_4) according to ISO 9614-1 [3] was performed to evaluate the accuracy of the measurements. Favorable results were obtained for the engineering degree of accuracy in all the cases.

The base partition element consists of a laminated plasterboard wall built on metal profiles. Even though there are other eligible elements such as brick or plaster walls, the laminated plasterboard wall is proposed as a base reference due to the extensive bibliography available for such element [28]. In addition, this element is

Figure 4.
Example of the SONAH measurement layout.

frequently used either independently or combined with doors as partition solutions, forming vertical walls between rooms.

Figure 5 shows the results through intensity and beamforming for the whole wall and through SONAH in three points of the wall in the third octave band. It can be observed, in the three images, that the wall is homogeneous separating element without leaks and, as a consequence, the intensity and pressure maps are uniform over all the surface of the element without leakages or fissures.

In the case of the anti-panic door, **Figure 6** shows the results obtained. **Figure 6(a)** shows the graphical results obtained from the sound intensity measurements (the upper-left point should not be considered because a measurement error took place). Analyzing the results obtained, it is possible to observe that the door exhibits a very homogeneous behavior with very little differences in sound intensity transmission and in addition in the sound insulation. Taken into account that, according to the intensity standard ISO-15186-2 [2], it is not possible to expand the measurement grid until the floor, it was not possible to identify the leakage in the bottom part of the door because the measurement grid does not cover this part.

The sound pressure maps obtained with beamforming and SONAH are those shown in **Figure 6(b)** and **(c)**, where it is possible to observe the areas where there is a higher sound pressure level emitted, in other words, the areas with the lower sound insulation. This area is located at the bottom of the door, because there is no adjustment between the door and the floor. The rest of the door exhibits a very homogeneous behavior because there is a big difference in sound pressure level among the main noise source and the rest ones.

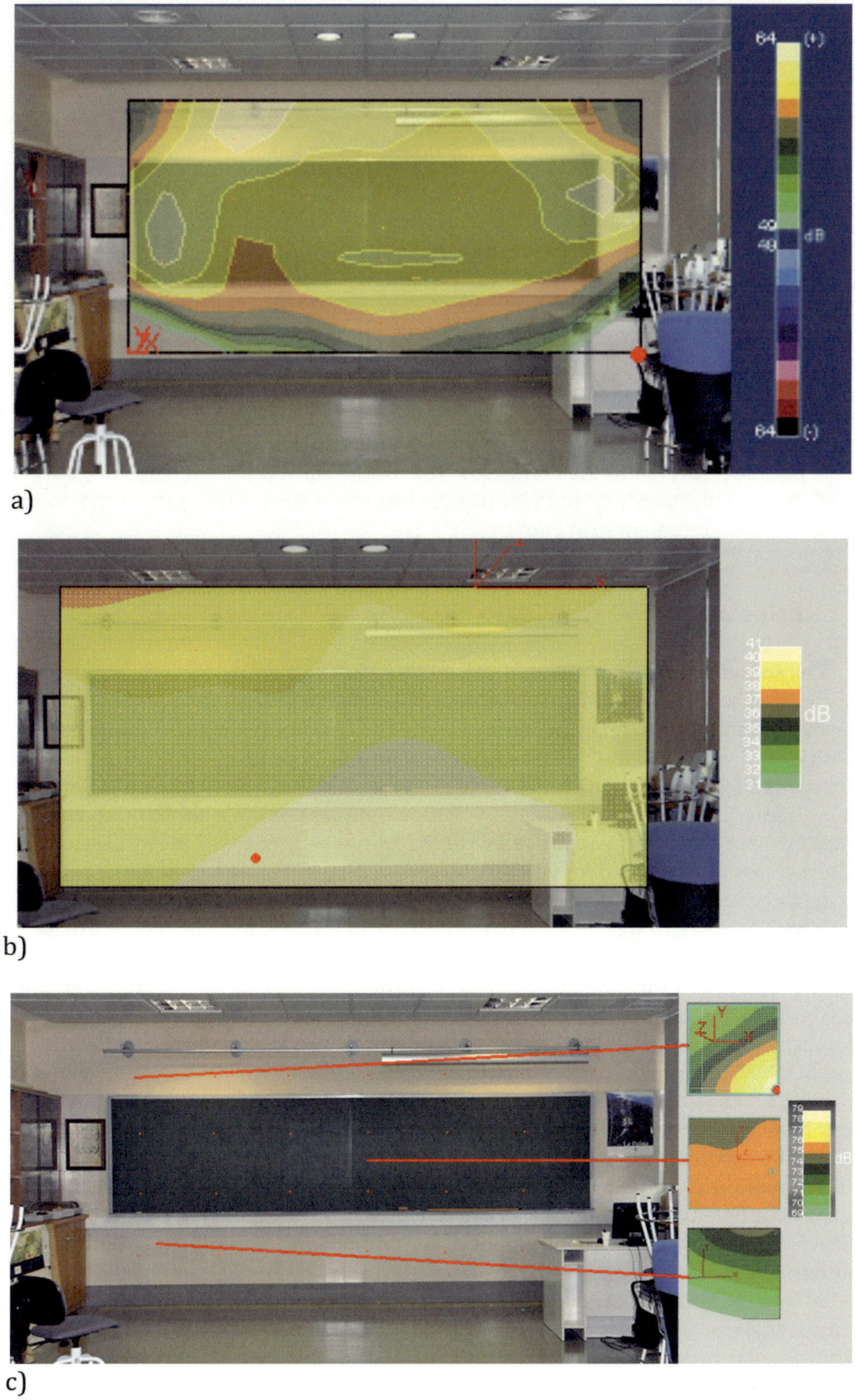

Figure 5.
Results obtained for the base element. (a) Intensity map (200–5 kHz). (b) Pressure map with Beamforming (1-5 kHz) with 10 dB dynamic range. (c) Pressure map with SONAH (160-1600 Hz) with 10 dB dynamic range.

a) b) c)

Figure 6.
Results obtained for the anti-panic door. (a) Sound intensity map (200–5 kHz), (b) pressure map with beamforming (1–5 kHz) with 10 dB dynamic range, and (c) pressure map with SONAH in the door center (160–1600 Hz) with 10 dB dynamic range.

If both measurements are compared, the results show a similar behavior, but intensity measurements do not identify the leakage at the bottom part of the door due to the measurement grid chosen.

3. Conclusions

The sound pressure technique based on ISO 16283-1 regulation only requires an analyzer with a third-of-an-octave accuracy, both in sound pressure level and reverberation time readings. In order to apply the measurement technique based on ISO 15186-2, however, it is necessary to use an intensity probe kit in conjunction with a two-channel analyzer with higher sensitivity than that required for sound pressure measurements. Beamforming and SONAH techniques require complex and costly sound instrumentation if compared with that commonly used for in situ assessment of airborne sound insulation, whether by pressure or intensity techniques; this is due to the large number of channels required, the application of processing algorithms with high computational cost and to the design, and preparation of specific microphone arrays.

Although beamforming and SONAH techniques are usually successfully applied in free-field conditions, the use of these techniques under different conditions pinpoints those areas with higher sound transmission. The joint use of beamforming and SONAH techniques on the construction elements under study quickly allows to identify areas of maximum sound transmission on the constructive element under test. In addition, whereas the setting up and the verification of the measurement system based on beamforming and SONAH are initially a time-consuming task, subsequent measurements are carried out in less time than that needed for pressure and intensity techniques, resulting in a greater repeatability of the test and therefore with less influence of potential sound field variations during the in situ measurement.

In beamforming technique applications, the best results are obtained without reference signals, while in SONAH technique are obtained using the reference from accelerometer signals or from the central microphone of the array. The use of alternative processing algorithms in beamforming would allow greater detail and definition on the identification of different sound transmission areas through the partition element.

In conclusion, the implementation of a measurement system based on the combined use of beamforming and SONAH techniques identifies in an accurate way leakages and weakened areas of partition elements under test. This working model allows to analyze quickly and effectively the behavior of acoustic insulation in areas where there are leaks or weakenings of the partition elements since, once configured the instrumentation system, it is only necessary to perform one measure at each application distance of the corresponding technique to obtain a first estimation of the partition element sound insulation quality.

Acknowledgements

This methodology was developed and granted by the project "Determination of the acoustic insulation of solid, acoustic and fire-resistant doors" (JCCM, Ref. PPII10-0172-426).

The publication of this chapter was funded by Escuela Politécnica de Cuenca - UCLM.

Author details

José A. Ballesteros*, Samuel Quintana and Marcos D. Fernandez
Escuela Politécnica de Cuenca, Universidad de Castilla-La Mancha, Cuenca, Spain

*Address all correspondence to: josea.ballesteros@uclm.es

References

[1] International Organization for Standardization, ISO 16283-1:2015. Acoustics–Field Measurement of Sound Insulation in Buildings and of Building Elements—Part 1: Airborne Sound Insulation; 2015

[2] International Organization for Standardization, ISO 15186-2:2003. Acoustics. Measurement of Sound Insulation in Buildings and of Building Elements Using Sound Intensity. Part 2: Field Measurements; 2003

[3] International Organization for Standardization, ISO 9614-1:1993. Acoustics-Determination of Sound Power Levels of Noise Sources Using Sound Intensity—Part 1: Measurement at Discrete Points; 1993

[4] Ballesteros JA, Quintana S, Fernández MD, Martínez L. Application of beamforming and SONAH to airborne noise insulation measurements. In: Berlin Beamforming Conference; 2012

[5] Johnson DH, Dudgeon DE. Array Signal Processing: Concepts and Techniques. Englewood Cliffs: PTR Prentice Hall; 1993: pp. 1-523

[6] Hald J. Basic theory and properties of statistically optimized near-field acoustical holography. The Journal of the Acoustical Society of America. 2009;**125**:2105-2120

[7] Lafon B, Antoni J, Sidahmed M, Polac L. Cyclic sound intensity and source separation from nah measurements on a diesel engine. Journal of the Acoustical Society of America. 2008;**123**(5):3387

[8] Gerges SN, Fonseca WD, Dougherty RP. State of the art beamforming software and hardware for applications. In: Proceedings of the 16th International Congress on Sound and Vibration; 2009

[9] Nordborg A, Wedemann J, Willenbrink L. Optimum array microphone configuration. In: INTER-NOISE and NOISE-CON Congress and Conference Proceedings, no. 6, Institute of Noise Control Engineering; 2000. pp. 2318-2323

[10] Marroquin M. A comparison of seven different noise identification techniques. SAE Transactions. 2003;**112**(6):2141-2151

[11] Hald J, Christensen J. A class of optimal broadband phased array geometries designed for easy construction. In: INTER-NOISE and NOISECON Congress and Conference Proceedings, no. 3, Institute of Noise Control Engineering; 2002. pp. 1993-1998

[12] Christensen J, Hald J. Technical Review, Beamforming, B&K Technical Review 1; 2004

[13] Ginn K, Hald J. Combined NAH and beamforming using the same microphone array. In: Forum Acusticum; 2005

[14] Ginn K, Hald J, Gade S. STSF-Practical instrumentation and application. Digital filter analysis: Real-time and non real-time performance. B & K Technical Review. 1989;**2**:21-27

[15] Capon J. High-resolution frequency-wave number spectrum analysis. Proceedings of the IEEE. 1969;**57**(8):1408-1418

[16] Sijtsma P. Clean based on spatial source coherence. International Journal of Aeroacoustics. 2007;**6**(4):357-374

[17] Brooks TF, Humphreys WM. A deconvolution approach for the mapping of acoustic sources (DAMAS) determined from phased microphone

arrays. Journal of Sound and Vibration. 2006;**294**(4):856-879

[18] Sarradj E, C Schulze C, Zeibig A. Identification of Noise Source Mechanisms Using Orthogonal Beamforming. Noise and Vibration: Emerging Methods; 2005

[19] Schmidt R. Multiple emitter location and signal parameter estimation. IEEE Transactions on Antennas and Propagation. 1986;**34**(3):276-280

[20] Sarradj E. A fast signal subspace approach for the determination of absolute levels from phased microphone array measurements. Journal of Sound and Vibration. 2010;**329**(9):1553-1569

[21] Hald J. Patch near-field acoustical holography using a new statistically optimal method. B & K Technical Review. 2003;**1**:40-50

[22] Dirks G, Gade S, Hald J, The use of near field acoustical holography for leak. In: INTER-NOISE and NOISE-CON Congress and Conference Proceedings, Institute of Noise Control Engineering; 2005

[23] Instituto Eduardo Torroja de ciencias de la construcción (Eduardo Torroja Institute for Building Sciences), Código Técnico de la Edificación-CTE; Catálogo de elementos constructivos del CTE (Technical Code for Building-CTE; Catalogue of building elements of the CTE); 2010

[24] Asociación Española de Normalización (Spanish Association for Normalization), UNE-EN 12354-1:2000. Estimación de las características acústicas de las edificaciones a partir de las características de sus elementos. Parte 1: Aislamiento acústico del ruido aéreo entre recintos. (Estimation of the acoustic characteristics of buildings from the characteristics of their elements part 1: Airborne sound isolation between rooms); 2000

[25] Brüel & Kjaer. LAN-XI data acquisition hardware for PULSE and test for I-deas. Brüel & Kjaer Product Data. 2016:1-40

[26] Brüel & Kjaer, PULSETM Array-based Noise Source Identification Solutions: Beamforming Type 8608, Acoustic Holography Type 8607 and Spherical Beamforming Type 8606, Brüel & Kjaer Product Data 1-12; 2013

[27] Jacobsen F, Tiana-Roig E. Measurement of the sound power incident on the walls of a reverberation room with near field acoustic holography. Acta Acustica United with Acustica. 2010;**96**(1):76-81

[28] Ortí JS, Galiana JL, Martínez AU. Aislamiento acústico de particiones ligeras multicapa: estudio del material absorbente y del sistema de montaje (Sound isolation of light multilayer partitions: Study of the absorbent materials and the assembly system). Universidad Politécnica de Valencia; 2003